AF700539

Subnature

LISA
♡'s
MIKE

B_mu

PUDDLES

MATTER

LIFE
TAFT
WISCONSIN
KANSAS
IKE

ATMOSPHERES MATTER LIFE

David Gissen

Subnature

ARCHITECTURE'S OTHER ENVIRONMENTS

ATMOSPHERES

MATTER

LIFE

Princeton Architectural Press
New York

Published by
Princeton Architectural Press
37 East 7th Street, New York, NY 10003

For a free catalog of books, call 1-800-722-6657
Visit our website at www.papress.com

© 2009 David Gissen
All rights reserved
Printed and bound in China
12 11 10 09 4 3 2 1 First edition

The author gratefully acknowledges the generous support of the Chalsty Research Award, part of the Chalsty Aesthetics and Philosophy Initiative, made possible by a grant from the Chalsty Foundation.

No part of this book may be used or reproduced in any manner without written permission from the publisher, except in the context of reviews.

Every reasonable attempt has been made to identify owners of copyright. Errors or omissions will be corrected in subsequent editions.

Editor: Laurie Manfra
Designer: Paul Wagner

Special thanks to: Nettie Aljian, Bree Anne Apperley, Sara Bader, Nicola Bednarek, Janet Behning, Becca Casbon, Carina Cha, Penny (Yuen Pik) Chu, Carolyn Deuschle, Russell Fernandez, Pete Fitzpatrick, Wendy Fuller, Jan Haux, Clare Jacobson, Aileen Kwun, Nancy Eklund Later, Linda Lee, John Myers, Katharine Myers, Lauren Nelson Packard, Dan Simon, Andrew Stepanian, Jennifer Thompson, Joseph Weston, and Deb Wood of Princeton Architectural Press
—Kevin C. Lippert, publisher

Library of Congress Cataloging-in-Publication Data
Gissen, David.
Subnature : architecture's other environments / David Gissen.
—1st ed.
p. cm.
Includes bibliographical references and index.
ISBN 978-1-56898-777-4
1. Architecture—Environmental aspects. 2. Architecture—Human factors. 3. Buildings—Environmental engineering.
I. Title. II. Title: Architecture's other environments.
NA2542.35.G57 2009
710'.47—dc22

2009008062

CONTENTS

Subnature was realized with the inspiration, commentary, and support of numerous people. First, I would like to thank the committee for the Chalsty Research Award at the California College of the Arts (CCA), and the award's benefactor, whose incredibly generous grant enabled the realization of this book. Second, as I finish this project and begin transforming my dissertation into a book, I must thank my advisors Matthew Gandy and Adrian Forty for steering me deep into the history and theory of nature-space dynamics. I am sure that they are as surprised as I am that the literature they recommended and architectural work they wanted me to understand could come together in this particular form. My fellow PhD space-nature traveler, Ben Campkin, has been tremendously supportive, and I both refer to his research and pay particular attention to steer clear of his forthcoming work on concepts of filth and urban development. Colleagues at CCA have both inspired and warmly challenged me in ways that most academics rarely enjoy. Good friends that either discussed the material with me or provided welcome respites from work-related discussion include Mitchell Schwarzer, Andrew Kudless, Jordan Geiger, Miriam Paeselack, Thom Faulders, Anna Reiner, C. Greig Chrysler, Hugh Pocock, Javier Arbona, Federico Windhausen, Joseph Tanke, Molly Slota, and Patrick Wright. Friends and editors who read chapters include Ronald Rael, Ray Ryan, and Thomas Weaver. At Princeton Architectural Press, I would like to thank Clare Jacobson for approaching me about writing a new book and editor Laurie Manfra for discussing the book with me and helping me transform the following chapters into a more carefully crafted argument. Nick Brown provided assistance gathering images from the contemporary architects included in this book. I would like to thank my family, the Dreyfuses, Porters, Schreibers, Mayers, and Schulmans, and particularly Linda Gissen, my mother, who has patiently endured my work schedule. This book is dedicated to my father, Ira Gissen, who passed away in 2005; he is sorely missed by his family and friends. My wife Rachel Schreiber is on my mind as I complete this project. When we first met, Rachel reopened my eyes to the pleasures of intellectual inquiry; her friendship, humor, inspiration, and intimate companionship over the past eight years is one of the greatest pleasures of my life.

Five years ago I began a PhD on the subject of architecture and nature, but like most doctoral students, I was not entirely certain what I would explore. When I began, my advisors suggested that I investigate contemporary literature for existing texts and images related to the concepts of nature in both architecture and urbanism. They encouraged me to search far and wide, to dredge historical and contemporary architectural, urban, and geographic literature, exhibitions, and websites. The results of mining these secondary sources were collected in files. Categorizing what I had encountered, I found myself drawn to many peculiar things, much of it overlooked in the more general discussion of what might be termed *natural architecture*. Such hoarding of information typifies the earliest stages of any research project, and these efforts are usually pitched and forgotten as research and a serious thesis take shape. When I was finishing my dissertation (an exploration of nature in modern buildings in New York City in the 1970s), I received a letter from Princeton Architectural Press inviting me to develop a new book on nature in architecture. Initially I assumed that I would turn down this opportunity and instead focus on publishing my dissertation, but I soon realized that my collection of quotes, images, and contemporary projects on the subject of nature and architecture was a modest treasure. I had amassed a considerable amount of information that, as a whole, is simply not discussed in contemporary debates on the intersections of architecture and nature. I realized that this material on smoke, exhaust, dust, the heat of crowds, and mud addressed a type of nature that was wholly undertheorized, underdiscussed, and undervisualized in architecture. It seemed that my secondary material was, in fact, worthy of revisiting. I decided to do three things: first, I wrote to Princeton Architectural Press, explaining that rather than pursue another book on "nature," they might consider this other, stranger form of nature that I had been exploring; second, I added to my existing collection of material additional studies of forms built upon the sources I had already explored; and third, I decided to take a closer look at how these alternative natural forms appear in contemporary architectural practice. The sum of this material is gathered here.

This book explores contemporary and historical imagery and writings, identifying a new material-aesthetic between architecture and nature. I draw on architectural and urban design theorists' key texts and contemporary practitioners' recent designs to examine how both groups envision peripheral and often denigrated forms of nature, which I refer to as "subnature." I argue that, forms of nature become subnatural when they are envisioned as threatening to inhabitants or to the material formations and ideas that constitute architecture. Subnatures are those forms of nature deemed primitive (mud and dankness), filthy (smoke, dust, and exhaust), fearsome (gas or debris), or uncontrollable (weeds, insects, and pigeons). We can contrast these subnatures to those seemingly central and desirable forms of nature—e.g., the sun, clouds, trees, and wind. These latter forces are generally worked into the forms, practices, and ideas that constitute the primary realization of nature within architecture. While several books examine the historical engagement between architecture and these more normative forms of nature, this book is the first to uncover the story of the subnatural in architecture.[1] It examines how a wide range of architects and theorists identify lesser forms of nature and the complex ways that they engage with them.

While most practitioners viewed architecture as a refuge from subnature, a few modern and many more contemporary architects examined how architecture might negotiate a rapprochement with subnature. They reached out to forms of subnatural life, such as pigeons, as metaphors for a socially liberated architecture or considered how buildings might literally incorporate puddles and other inundations of water into their forms. More radically, a few contemporary architects imagine how architecture might produce subnatural forms and experiences, envisioning buildings as places in which formerly desirable plants suddenly appear as invasive weeds, or in which buildings release the dankness of the earth into the spaces of a home. As this book will demonstrate, the story of subnature is deep and complex. Nothing is inherently subnatural in architecture; rather, these forms arise from a particular set of social and architectural practices. Nevertheless, I argue that any approach relating architecture to nature must acknowledge this "other" nature embedded within the history of architectural thought, appearing in aggressive ways throughout the most progressive contemporary practices.

Aspects of what I term *subnature* have been partially labeled by recent historians and theorists. They are found in what Antoine Picon refers to as *anxious landscapes* and what the landscape architect Gilles Clément terms the *third landscape*. Both terms often describe postindustrial spaces at the peripheries of cities, where rusting buildings, weeds, and industrial debris coalesce. It may be identified in what architect François Roche terms *corrupted biotopes* to describe, among other things, the polluted water and air that produces hermaphroditic fish and polar bears.[2] These authors ask that we consider how a more radical concept of nature might enable us to imagine a significantly modified architectural and social milieu. I argue that we might more aptly identify the forms of nature identified by these authors—and the additional forms collected in this book—as subnature, as they are positioned below other forms of nature. If the supernatural is a world of miracles, a religious world above nature, and the natural is the world in which human society is located, then the subnatural is the realm in which we can barely exist in the state that we currently conceive ourselves, both socially or biologically.[3] It is that zone that is most fearsome, because it describes the limits in which contemporary life might be staged. It is thus no coincidence that subnatures are generally marginalized in architecture. When they appear in architectural thought in a nonmarginal way, they are often used to describe the passage of societies: in drawings of weeds, puddles, and dusty ruins by Giovanni Battista Piranesi and Robert Adam; in sketches of debris from the ravages and recovery from war by Alison and Peter Smithson and Arata Isozaki; or the possibilities for something more or less than human, in images of people, insects, and animals cohabitating by François Roche, Efran Garcia Grinda, and Cristina Diaz Moreno. But the subnatural is not the apocalyptic edge of society. Rather, it reveals another possible form of nature in which we can be something more or less than is currently possible within our conceptions of nature.

I believe that an understanding of subnature is a crucial addition to contemporary debates regarding architecture, urbanism, and nature. *Subnature: Architecture's Other Environments* offers an alternative vision to those contemporary municipalities, developers, and architects who seek to remake cities and buildings through the parameters of a more natural framework based on sustainable principles. *Subnature* also offers an alternative to the emerging vitalist discourse on "flow" as the dominant effect of nature in architecture. The recent green movement has advanced a laudable critique of contemporary building practices. In numerous books, we have learned to engage sunlight, climatic systems, and the wind currents that stretch across the Earth's surface.[4] While these strategies address energy conservation and foster an appreciation of nature, as employed, they advance a seemingly neo-Victorian and neo-Haussmannite vision of urbanism in many global cities. I call this approach neo-Victorian because, like the reformers of the nineteenth century, green architecture often entails the utilization of nature as an instrument that cleans the world, increases productivity and efficiency, and transforms our existing natural relationship, while advancing the social sphere as it exists. I consider these approaches neo-Haussmannite because, like the remaking of Parisian space under Georges-Eugène Haussmann, the introduction of green building often enhances the power of urban wealth in the name of mending a natural relationship.

I do not think most "green" architects wish this to be so, but one only has to consider the recent green building booms in New York's Times Square and Battery Park City, and the corresponding production of a homogenous and elite social sphere, to understand how the restoration of nature is used to re-establish a specific class-based idea of the city. The above criticism of green architecture may also be slightly redirected to critique the transformation of human experience with respect to related parameters of *fields* and *emergence*. Antoine Picon has already noted that what is often termed *field theory*, *self-organization*, or *emergence* often relies on a "strange, vitalistic conception of the world."[5] But we might also add that more explicit theories of dynamic flow view nature as little more than a circulatory construct whose role is to erase all of the stagnancies that stand in its way. Nothing could provide a more apt metaphor for the recent rebuilding of cities for a global society on the move.[6]

Subnatures are the very natures that stand in contrast to the above processes. In comparison to the demographic distribution of green architecture, what I identify as subnatures are primarily experienced as aspects of the seemingly subhuman conditions of contemporary urbanization and its subcultural peripheries. They are also those natures that stand against the remaking of the world into a pulsing circulatory apparatus. Like the processes described above, these relationships extend back to the nineteenth century. For example, if we consider documentary literature of nineteenth-century England, we see how—in Karl Marx's passages on English working conditions, in Friedrich Engel's accounts of Manchester, in Henry Mayhew's accounts of "mudlarks," or in Gustave Doré's drawings of the London gasworkers—a seemingly unnatural and often stagnant nature is navigated by those at the margins of the urban social structure.[7] In these accounts, dust, dirt, puddles, and vermin are the repulsive, murky, frictional natures of industrial urban life. Perhaps more interestingly, subnature is also wielded by those at the margins as a form of social change. If we shift from London to Paris, we see how, in 1848 and 1871, revolutionaries erected barricades to block the use of streets (as conduits) with the socionatural detritus of industrial urbanization: mud, garbage, debris, and even animal carcasses. The statement made at the barricades in 1968, 120 years after these early Parisian riots—"under the street, the beach"—and oft-quoted by neo-avant-garde architects in the early 1970s is about the power of social and physical transformation and the corresponding, strange appearance of subnature. In recent subcultural urban movements, images of mud, bones, smoke, and exhaust are wielded as provocations within the contemporary urban sphere. Individuals engage strategically with subnature, forming para-urban social networks, as with pigeon breeders or weed tillers working on abandoned urban sites. Subnature will not save us from our inequities, but its inherently alienating character enables us to consider how more comforting forms and dynamic images of nature are often used to reproduce existing forms of power in society.[8] Ultimately, we can marginalize the subnatural in architecture in the name of the natural, or we might consider the possibilities of exploiting subnature as a form of agitation or intellectual provocation.

Subnature is a book that advances ideas about denigrated forms of nature from a predominantly theoretical and historical vantage point. It differs from recent books—such as *Earth Architecture*, or more established books such as *The Granite Garden*—that recuperate the hidden technological sophistication or unrecognized beauty of underappreciated forms of nature, such as mud and weeds.[9] These works explore how architects and landscape architects might stake out a new appreciation of these forms based on economy, technology, and environmental health. In contrast, *Subnature* focuses on the historical circumstances that make recoveries of mud and weeds interesting and aesthetically powerful. I argue that the power of modern earth buildings or weed-infested landscapes draws from historical ideas and images about these forms of nature. Furthermore, many of the architects of the contemporary projects in this book have a pronounced and deep relationship with the particular history of denigration that moves through architectural engagements with mud, weeds, dankness, or debris—a historical awareness expressed in their architecture and writings.

In pointing out the theoretical and historical dimensions of this book, I would also like to note that the concept of subnature differs from the relatively recent theories of weathering advanced by authors such as Mohsen Mostafavi and David Leatherbarrow.[10] The theories proposed by these authors have resulted in a recent appreciation of materials that decay or show the stains that architecture accumulates over time. Their theories of weathering opened a profound new critique in which modern architecture's resistance to the surrounding elements appeared questionable. In proposing a critical new concept of buildings that are open to the elements, Mostafavi and Leatherbarrow advanced a curiously uncritical conceptualization of the environment. Within these theories of weathering, environments appear as fixed and stable systems relative to a dramatically changing architectural object. This concept has influenced a whole host of building practices in which architects have "revealed" rain, air, or urban grit through patinas and staining. In contrast, subnature is not only about exposing the realities of external environments. Rather, the theory of subnature proposed within this book supports the notion that architecture and the environment are produced simultaneously. In fact, many of the projects in *Subnature* effectively invert the paradigm of weathering. They appear to recode the surrounding environment, infusing new concepts of nature and time into their surroundings. It is that latter aspect that adds to our often troubling experiences of subnature.

Admittedly, *Subnature* emerges from a complex intersection of concepts, but this book is intended to appeal to architectural students and emerging professionals. It contains twelve brief, heavily illustrated chapters conveying the potential for architecture and theory to be grounded in the concept of the subnatural. Each chapter provides an examination of a particular form, such as dust, debris, exhaust, and its actualization in contemporary practice. These are organized from the most atmospheric, dankness; to those found in life, crowds. In the text of each chapter, I uncover the concepts surrounding these subnatures in architectural thought, typically moving from the origin of modern Western architectural theory from the eighteenth century to the present. Some chapters concentrate on particular struggles or geographies where pronounced engagements with a particular form of subnature now resonates with contemporary architectural work. In illustrating these chapters, I chose a few key images from architectural history, complementing these with many more from architectural practices of the past fifteen years. Most chapters conclude with discussions of the most recent projects that, I believe, illustrate a particular form or relationship under examination. This format locates the book somewhere between an exhibition catalog and an architectural theory book. It makes these concepts accessible and useful for the student and professional, and it provides both a theoretical-historical backbone and a method for considering the ideas under review. This book could be the first of numerous efforts identifying the complexity and social power that emerges when we consider all of the peripheries of architecture-nature relations.

1. For recent books on the history, theory, and design concepts of natural architecture, see James Wines, *Green Architecture*, ed. Philip Jodidio (New York: Taschen, 2000); Sarah Bonnemaison and Christine Macy, *Architecture and Nature: Creating the American Landscape* (London: Routledge, 2003); Alessandro Roca, *Natural Architecture* (New York: Princeton Architectural Press, 2007).

2. Antoine Picon, "Anxious Landscapes: From the Ruin to Rust," *Grey Room* 1 (Fall 2000): 64-83; Gilles Clément and Philippe Rahm, *Environ(ne)ment: Approaches for Tomorrow* (Montreal: Canadian Centre of Architecture, 2006); François Roche, *DD 05R&Sie: Corrupted Biotopes* (Seoul: Damdi, 2005).

3. The relationship between the supernatural and subnatural is partially inspired by Robert Scholes, *Structuralism in Literature: An Introduction* (New Haven: Yale University Press, 1975), 119-20.

4. Consider G. Z. Brown and Mark DeKay, *Sun, Wind, and Light: Architectural Design Strategies* (New York: Wiley, 2001) or more theoretical texts such as William McDonough, "Design, Ecology, Ethics and the Making of Things," in *Theorizing a New Agenda for Architecture: An Anthology of Architectural Theory 1965-1995*, ed. Kate Nesbitt (New York: Princeton Architectural Press, 1996), 398-407; or the book that McDonough coauthored with Michael Braungart, *Cradle to Cradle: Remaking the Way We Make Things* (New York: North Point Press, 2002).

5. See Antoine Picon, "Architecture, Science, Technology, and the Virtual Realm," in *Architecture and the Sciences: Exchanging Metaphors*, ed. Antoine Picon and Alessandra Ponte(New York: Princeton Architectural Press), 307. For more on field theory, see Sanford Kwinter, "La Cittá Nuova: Modernity and Continuity," in *Architectural Theory: An Anthology from 1871 to 2005*, ed. Harry Francis Mallgrave and Christine Contandriopoulos (Malden, MA: Blackwell, 2008), 474-75; and Stanley Allen, "From Object to Field," *Architecture After Geometry, Architectural Design* (1998): 24-31.

6. I first heard a critique of this type during a lecture by Nils Norman at the Maryland Institute College of Art, Spring 2004. See Nils Norman, *The Contemporary Picturesque* (London: Book Works, 2000).

7. For accounts of these authors' impressions of what is termed here as *subnature*, see the various chapters in William Cohen and Ryan Johnson, eds., *Filth: Dirt, Disgust, and Modern Life* (Minneapolis: University of Minnesota Press 2005); and Lynda Nead, *Victorian Babylon: People, Streets, and Images in Nineteenth-Century London* (New Haven: Yale University Press, 2000).

8. In emphasizing these last points, I should mention that many of the subnatural historical and contemporary works in this book emerge directly from official, instantiated forms of power—museums, governments, and the wealthy patrons who often commission architects to design their residences. Nevertheless, I believe the potential of subnature is locked within the idea of producing forms of nature as instruments against the dominant appearance of spatial power.

9. Ronald Rael, *Earth Architecture* (New York: Princeton Architectural Press, 2008); and Anne Whiston Spirn, *The Granite Garden: Urban Nature and Human Design* (New York: Basic Books, 1985).

10. Mohsen Mostafavi and David Leatherbarrow, *On Weathering: The Life of Buildings in Time* (Cambridge, MA: MIT Press, 1997).

PART ONE

ATMOSPHERES

Dankness

Smoke

Gas

Exhaust

Dankness

We begin this book's exploration of the history, theory, and design of subnature in the dark, wet, and cool spaces that mark the origins of architecture. The negotiation of that which we might term *dankness* refers to a surprisingly large aspect of the visual, textual, and technological culture of modern architecture. It appears in early discussions of the origins of architecture, tracts on infrastructural improvements to cities, manifestos on modern construction systems, and as a growing aspect of a more environmentally attuned architecture. Architectural theorists once associated dankness with natural and artificial subterranean spaces, such as caves and grottoes, but in the nineteenth century it increasingly became associated with types of underground spaces brought into being by urbanization, like sewers and basements. In the nineteenth century, architects and urbanists sought to remove dank spaces from the city and countryside. But in more recent years, they have begun to embrace the qualities that characterize dankness, seeing them as signifying the repressed within modern architecture or as representing sites of uncharted and immense particularity. As we shall see, there has always been some debate as to the virtue of dankness in architecture; for some, it represents one of the vilest types of architectural atmosphere; for others it offers a unique and underexplored spatial milieu.

From Vitruvius onward, architectural commentators and theorists viewed dark and wet caves as one of the earliest and fundamentally deficient spaces of human habitation. Vitruvius saw them as the primitive and wild habitats of early humans. Architecture for Vitruvius was precisely the art that provided an environment distinct from those of caves and other natural dwellings. French architectural theorist Marc-Antoine Laugier wrote of the cave as an inferior space of early habitation for primitive man—the "darkness and foul air make his stay unbearable."[1] For Laugier, the cave and architecture are opposed, as the flight from the dank space of the cave led to the construction of the first form of architecture, the "primitive dwelling." French architectural theorist A. C. Quatremère de Quincy viewed the cave somewhat more positively. He saw this dark subterranean space as one of three original types upon which all subsequent architecture was based; the other two were the tent and the hut. But, like other architectural theorists, Quatremère de Quincy believed that the cave was the most inadequate of his three types and the most resistant to architectural development. Finally, the dark, cool, and wet atmosphere of the cave served as a more explicitly metaphorical model in architecture. French classicist Philibert de l'Orme believed that the cave

Fig. 1.1
The builder emerging from a dark cave to become an architect, Philibert de l'Orme, 1568

was a metaphorical space of ignorance, and he drew an image of an architect emerging from a cave as a symbol of the profession. [**Fig. 1.1**] This latter concept of the cave draws on Plato, who famously wrote of the cave as a site where all knowledge is suspended in a representational world of shadowlike images.[2]

Despite the intense denigration of the cave in much architectural theory, this type of space, where darkness and moisture intermingle, was curiously embraced in several other forms of architectural thought and practice. The grotto—essentially a dark cavelike space that incorporated dripping fountains—became a central aspect of architectural thinking from the Renaissance to early modernity. Grottoes often featured quasinaturalistic and mythic scenes of underground worlds with technically sophisticated waterworks. They also invoked ideas of alchemic science—in which darkness, light, water, and earth combined. Also significantly, grottoes featured a specific type of atmosphere; within Italy and England, the dank climate of the grotto became associated with the production of a pleasing comfort. During Italy's hot summer months, grottoes were valorized for their cool and stable temperatures. In many ways, these spaces bridged the premodern and modern worlds by combining darkness and rustic forms with water technologies. A grotto developed by Jean-Jacques Lequeu for a Parisian park illustrates this concept. Lequeu included numerous waterfalls and a central pool in a quasisubterranean space that contained statuary evoking ancient mythology. In describing this mysterious space, Lequeu argued that it would offer "purification" to the user through an immersion in its waters. Such a claim relates to ancient mythological imagery, but it also engages emerging concepts of urban water use for hygiene. Lequeu's project is a harbinger for the gradual reworking of the city's water systems and underground spaces.[3] [**Fig. 1.2**]

Conflicting perceptions of dank spaces in early modernity gradually disappeared within the nineteenth century. Modern architects and urbanists warred against the dark, wet, and tepid spaces associated with the undergrounds of cities, most notably sewers and cellars. These previously mysterious areas within cities were now associated more explicitly with disease and vice. In the urban underground, air stagnated, moisture collected, and crime fomented. Within the context of modern architectural theory, Georges-Eugène Haussmann and Le Corbusier stand apart as two critical assessors of the modern city's underground regions. Haussmann charted and expanded the existing Parisian sewer system. Curiously, his work was an extension of earlier efforts by de L'Orme. In the sixteenth century, de L'Orme was commissioned by Henry II to explore Paris's medieval sewers but failed due to the cavernous and disgusting nature of the medieval sewerage system. In 1802, Napoleon Bonaparte commissioned Pierre-Emmanuel Bruneseau to chart the sewer system, and he succeeded after much effort. Bruneseau's maps were studied by Haussmann, who rebuilt and constructed new gas-lit, widened, and ventilated sewers as part of his immense transformation of the city.[4] [**Fig. 1.3**] Haussmann's sewers turned what was previously one of the most notoriously vile spaces in the city into a well-illuminated and ventilated underground system. In the early twentieth century, Le Corbusier, who admired Haussmann, further developed this transformation of the urban underground as part of an examination of the domestic cellar. Le Corbusier viewed the cellar as a type of dark, wet, unhygienic, and outmoded space that emerged from masonry bearing-wall construction. [**Fig. 1.4**] Writing of this construction system and its dank spaces, he proclaimed:

> Before concrete and iron, to build a house in masonry, wide trenches were dug in the ground to find a firm soil for the foundations. As the earth along the trenches would slide down, it was easier to remove the earth between them. Thus were created cellars, mediocre spaces, dark or poorly lit and generally damp.[5]

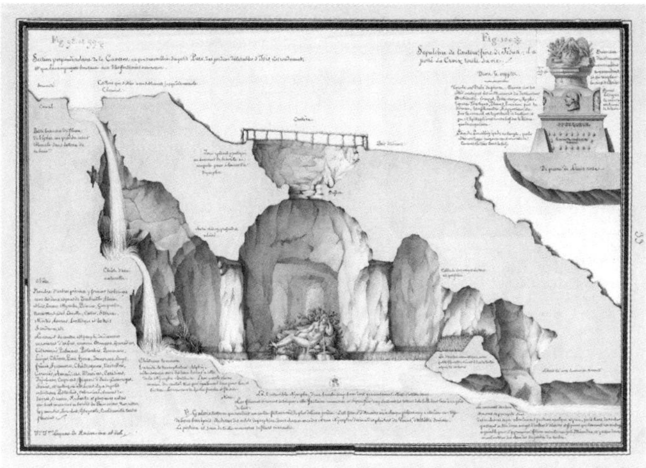

Fig. 1.2
The mythological character of the grotto, section of a proposed cave structure by Jean-Jacques Lequeu, 1792

Fig. 1.3
Rationalizing the dark, wet, underground spaces of Paris, interior of a sewer, Nadar, circa 1860

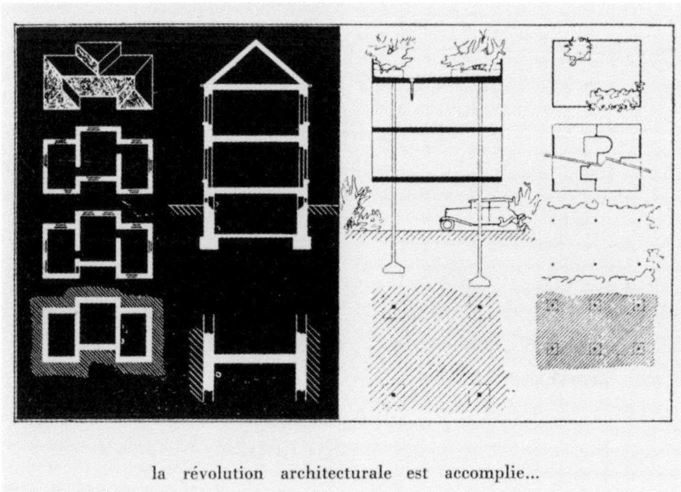

Fig. 1.4
The basement as countermodern space, in Le Corbusier's Five Points diagram, reprinted from *La Ville Radieuse* (*The Radiant City*), 1935

Le Corbusier's Five Points, as they would eventually be termed, sought to employ concrete and industrialized building materials to produce new forms of space and domesticity, effectively banishing the cellar (and the attic) from residential architecture. He wrote of the differences:

> With reinforced concrete you get rid of walls completely. Floors are carried on thin columns spaced far apart. To found them, a small well is dug out for each one down to good soil. Then the column is raised above ground. And at this moment one takes advantage of the situation. I don't need to take away this inevitable mass of earth in the heart of the house. My ground is intact, unbroken.... I draw an automobile on this regained ground, and I let air and vegetation go through.[6]

In a drawing comparing his new system to older construction technologies, Le Corbusier compared the *salubrité* of his concrete construction method to the damp and dark spaces of earlier masonry dwellings. Within modern architecture and planning, the dark, wet, and cold spaces of the urban underground had to go.

The clean-up of the urban sewer and the banishment of the cellar from modernity represented a victory of the rhetoric of light and air over the dark, the tepid, and the dank. But some midcentury theorists viewed the loss of this quality as a loss of history and the sanitization of domesticity as a form of repression. Postwar discussions of the dark spaces of premodernity engaged more with psychology than with the hygienic science examined by Haussmann and Le Corbusier. Reacting to housing schemes proposed by Le Corbusier and other modern urbanists, the French philosopher Gaston Bachelard sought a conceptual recovery of the basement or "cellar" within modern consciousness. Inspired by Carl Jung's concept of psychological archetypes, Bachelard believed that our understanding of space was fundamentally rooted in the experience of home that each of us carries with us in our daydreams and random thoughts. Modern space lacked the primordial aspects of our domestic dreams. Bachelard developed a type of science of the oneiric or dreamlike house, believing that our images of home were situated precisely between the two spaces banished by Le Corbusier: the cellar and attic. The attic spoke of a house's climate—including snow, rain, and wind—it is the more rational space of the house; while the cellar, he wrote, "is first and foremost the dark entity of the house, the one that partakes of subterranean forces. When we dream there, we are in harmony with the irrationality of its depths."[7] Bachelard wrote that the cellar was a space of "fear" and "buried madness." With the weight of the earth behind it, dark spaces and stagnant water reveal how the apparent autonomy of the house is haunted by links to primordial waters and other underground subnatures. Bachelard's influence within architecture was broad. In concert with ideas from philosopher Martin Heidegger, he instilled a return to concepts of place over space, and by the mid-1970s this involved a far more sophisticated exploration of metaphor and physical sensation within contemporary architecture.

A second type of exploration of a house's underground also relied on psychoanalytical theory but from a less Jungian and more explicitly Freudian-inspired understanding of the *unheimlich* or unhomely. This exploration of the underground wrestled with the modern home's apparent autonomy over the psychological trauma of modern existence. This approach, represented by the work of architect Peter Eisenman and historian Anthony Vidler, interrogated the home's false promise of stability—its groundedness. In Eisenman's work, this exploration initially entailed the manipulation of building components, such as walls and columns, held in tension against programmatic elements. But by the late-1970s, Eisenman pushed this exploration of the unhomely further by exploring an image of the stable "ground" of the house. In the Cannaregio project (1978), he situated a version of House Eleven in a large "excavation," as he termed it, open to the elements above. [**Fig. 1.5**] Within this dark "tomb," as it was labeled by Vidler, Eisenman construed a series of building volumes that themselves appeared as excavations.[8] Much has been made of the links between the operations of Eisenman and psychoanalytical and poststructuralist theory. The excavated ground represents "the presence of absence," but in addition to understanding the particular philosophical strands that move through

Fig. 1.5
Excavated ground as uncanny space, Cannaregio proposal for Venice, Italy, by Peter Eisenman, 1978

Eisenman's late-1970s work, we might also connect these excavations with the larger postwar recovery of subterranean space that Eisenman's philosophical opposites, such as Bachelard, also explored.

In considering the role of dankness in recent contemporary work, it appears that architects are less taken with the particular psychological impact of the underground's "return." The underground might offer more than just a confrontational image against architecture's normative social roles. Rather, work by contemporary experimental architects, such as Philippe Rahm, Aranda\Lasch, and SeARCH, considers the particularities offered by its cool, dark, and moist qualities, repurposing the underground into an entirely new milieu in which architecture is conceived and experienced. This work certainly relies on the unfamiliarity of the underground explored by Bachelard, Eisenman, and others, but such unfamiliarity is directly linked to the potential new forms of sensation. For example, in the Underground Houses (2005), designed for the French region of Vassivière, Rahm envisioned how the material qualities of subterranean space might transform our concept of home. [**Figs. 1.6 + 1.7**] Rahm writes that the aim of the Underground Houses is to

> construct an architecture for which the climate is in profound relation with the Earth and the subsoil. The interior space, its chemical and physical nature, is here a product of the local soil, literally produced in the Earth. Heat, produced underground, baked with geothermal energy,

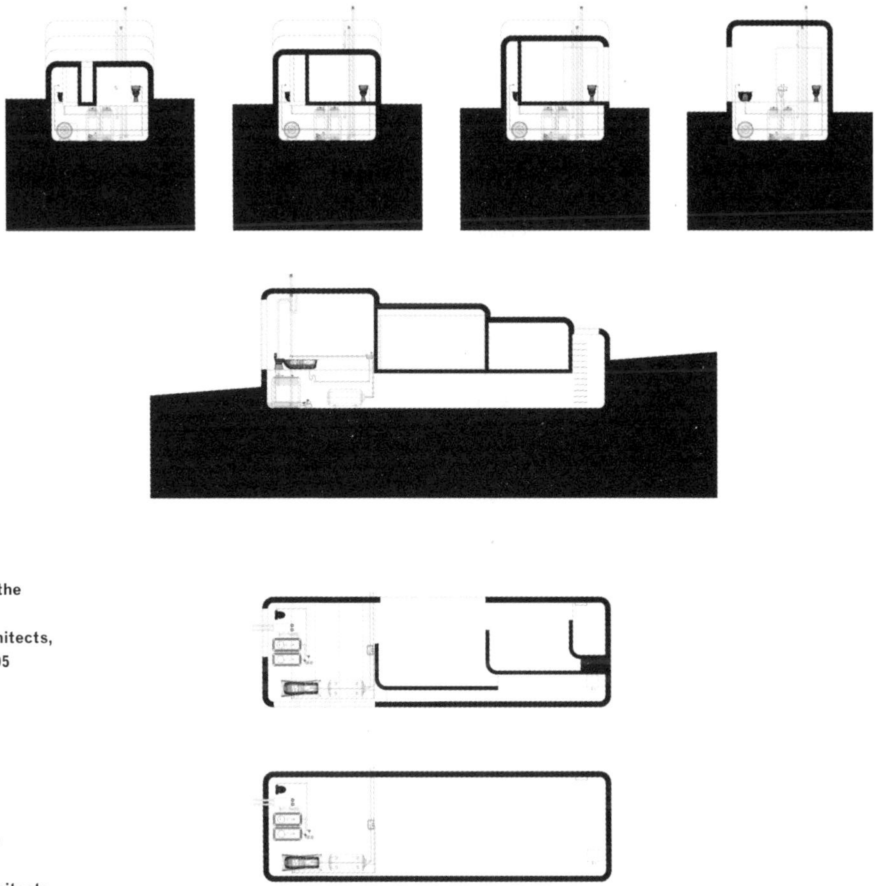

Fig. 1.6
Sections and plans of the Underground Houses, by Philippe Rahm Architects, Vassivière, France, 2005

Fig. 1.7
Aerial rendering of the Underground Houses, by Philippe Rahm Architects, Vassivière, France, 2005

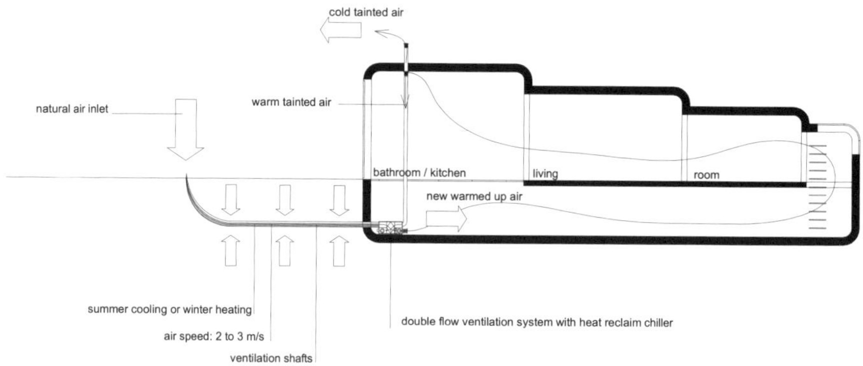

Fig. 1.8
Section diagram of climate system, Underground Houses, by Philippe Rahm Architects, Vassivière, France, 2005

Fig. 1.9
Perspective view of the Underground Houses, by Philippe Rahm Architects, Vassivière, France, 2005

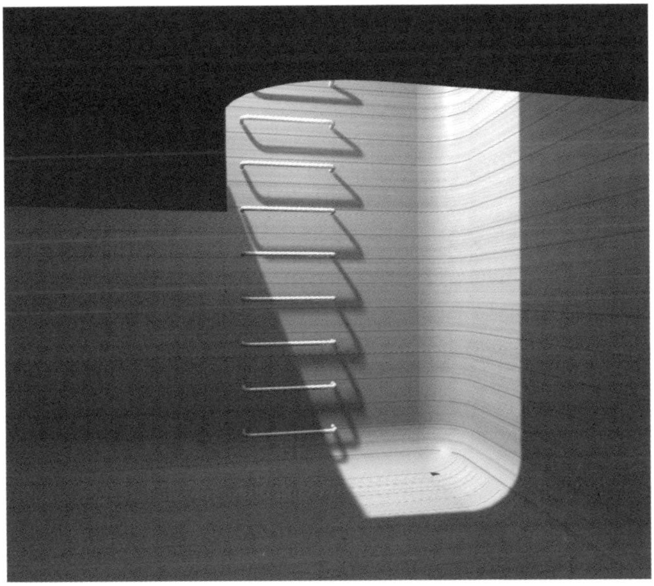

Fig. 1.10
Descending into the underground chamber, Underground Houses, by Philippe Rahm Architects, Vassivière, France, 2005

Fig. 1.11
View of the subterranean chamber, Underground Houses, by Philippe Rahm Architects, Vassivière, France, 2005

> air cooled within the cold darkness of the ground, the quality of a soil lived and breathed, the soil of Vassivière. A certain quality of air is sought above all, a certain freshness, drawn out from underground and stored in the inertia of the soil, as if we succeeded in giving the interior air of these houses an earthy taste, a slightly brownish tone.[9]

Each Underground House is organized as a two-story structure that sits nestled within a cavernous excavation—a trap for the atmosphere released by spaces within the earth. [Figs. 1.8 + 1.9] Rahm organized each space of the house relative to this cavern so that those rooms one might desire to be most open to moist, tepid air (the bathrooms) open to it, and those spaces that one might desire to be more isolated (the sleeping quarters) are largely protected from it. The underground chamber imparts a new type of strange zone within our image of domesticity. [Figs. 1.10 + 1.11] But Rahm's point is that once in touch with this uncanny space, we begin to understand it as an unactualized form of architecturally produced nature that reorients the house to the earth.

The above explorations of the underground continue in Villa Vals by SeARCH and Christian Mueller and in the Grotto project by Aranda\Lasch. Within Villa Vals the architects examine the very interface between above- and underground. They replace the craggy entrances of the formerly mysterious grotto with a pure platonic construct: a simple circle that makes the underground appear far less interstitial or alienating than those earlier explorations. The terraced underground house is oriented toward this circular opening. [Figs. 1.12 + 1.13] Architect Christian Mueller described the project as a "landscape project rather than a building," explaining that it produced an interior effect, like the Underground House, that he considered "weirdly comfortable"—an acknowledgement of the underground's conflicting image in late modernity. Finally, the architects Aranda\Lasch developed a project that explicitly returns us to the historical experience of those dark, cool underground spaces so prevalent in early-modern architectural thought. Grotto by Aranda\Lasch reconceptualizes the artifice of early explorations of underground spaces through contemporary modes of production, in a project constructed from digitally fabricated foam boulders. [Figs. 1.14 – 1.16] They write:

> The grotto is an artificial structure or excavation in a garden made to resemble a cave. It is always elaborately artificial, absurdly fake. Against this backdrop of theatricality, forbidden pleasures can occur: hidden and discovered, stolen and intimate. The grotto found its heyday in eighteenth-century English gardens, providing a dark and erotic

Fig. 1.12
Inhabiting the dark and cool regions of the Swiss Alps, perspective view of Villa Vals, by SeARCH and Christian Mueller, Vals, Switzerland, 2009

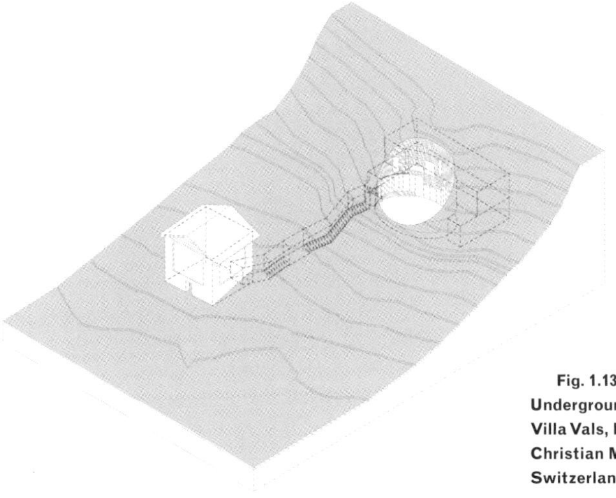

Fig. 1.13
Underground rendering of Villa Vals, by SeARCH and Christian Mueller, Vals, Switzerland, 2009

narrative to the landscape gardener's palette. This proposal for a temporary summer pavilion takes advantage of the grotto's essential feature: there is something to discover within, and that something is often wet.[10]

In all of these projects, the dankness of architecture is both reconceptualized and redesigned; it may still haunt space, but it also impacts new pleasures in our concepts of dwelling and home. Through these works, we realize what might pejoratively be termed *dankness* actually contains qualities relative to specific local materials or regions (a *terroir*, so to speak). These projects embrace the underground as a zone of expression, and they enable us to see that the seemingly undesirable quality of dankness can be filled with pleasure and that we may one day be able to recognize the underground's particular and varying features.

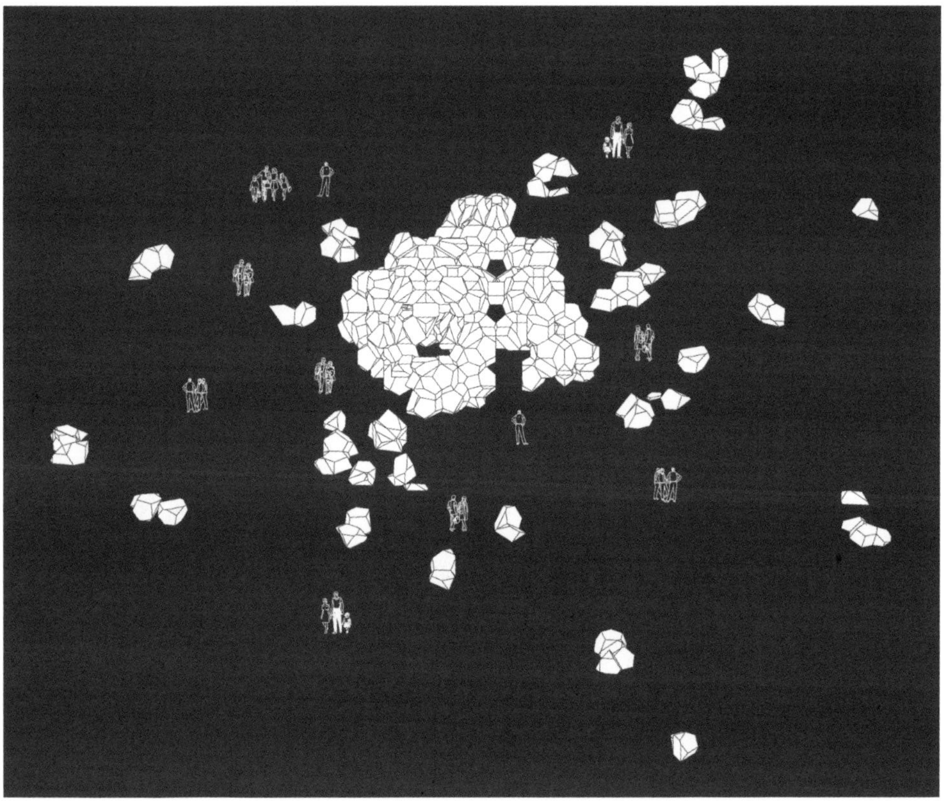

Fig. 1.14
Diagram of Grotto, by Aranda\Lasch, Queens, New York, 2006

Fig. 1.15
Model of Grotto,
by Aranda\Lasch, Queens,
New York, 2006

Fig. 1.16
Interior view of Grotto,
by Aranda\Lasch,
Queens, New York, 2006

Smoke

Ancient, early-modern, and modern architectural theorists speculate that social collectives first formed around fires. Vitruvius, whose writings are the oldest surviving texts on architecture, argued that the first aggregations of people took place around open fires created in forest clearings. His ideas were expanded in the lush Renaissance treatises of Antonio di Pietro Averlino and Cesare di Lorenzo Cesariano depicting entire societies centered around primordial fires. Hundreds of years later, during a period in which earlier classical theories of architecture were rejected, Gottfried Semper developed an architectural theory based on a belief that architecture began as a protective act surrounding the open-pit fire or hearth. Semper argued that primitive people developed earthworks, roofs, and wall coverings to help maintain the central hearth of their dwellings. Fire is a key aspect of architectural thinking on the origins of social collectives. Within these and other writings on architecture and fire, smoke is relegated to a lesser position and considered a nuisance. While Vitruvius mentions the significance of fire, his writings on smoke describe it as a potentially destructive presence in the home, something that soils and stains architecture. [**Fig. 2.1**] Semper wrote extensively about fire in his *Four Elements of Architecture* but did not incorporate smoke as a significant aspect of his theory. Reyner Banham, who offered one of the most evocative and contemporary theories of fire in his essay "A Home is Not a House," considers the smoky aspects of fire to be one of its key defects. Like other forms of subnature, smoke cannot be easily absorbed into architectural thought; when discussed, it is often visualized as strange and alienating. This is particularly true within architects' reactions to smoke-filled cities and spaces characterized by industrial modernization. We see this in architectural writings and imagery of the French Revolution, German idealist architectural theory, the English Arts and Crafts movement, and the Chicago School. In these various movements, smoke appears as strange emissions from chimneys and smokestacks and as exhalations within spaces through the very act of people smoking indoors. It is within these writings that we witness how smoke often intertwines ideas about bodies and buildings with additional concepts of class and health. Understanding this architectural relationship to smoke is important, as it enables us to develop a more considered role for this most unwanted form of subnature, especially in an era in which the sophistication and wealth of contemporary cities is defined by the absence of smoke.

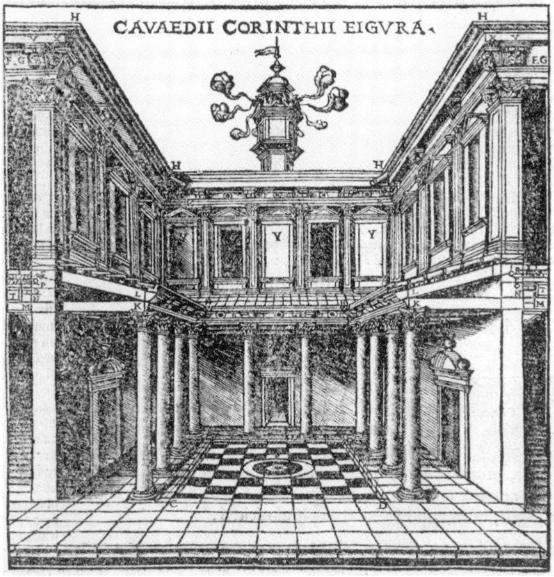

Fig. 2.1
A domestic interior emitting smoke, *The Roman House*, illustrated by Cesare di Lorenzo Cesariano, 1521

Fig. 2.2
The smoke-filled Black Country of the English Midlands, mid-19th century

Fig. 2.3
An allegorical image of the smoky Midlands, *Pegasus Above the City*, Karl Freidrich Schinkel, 1837

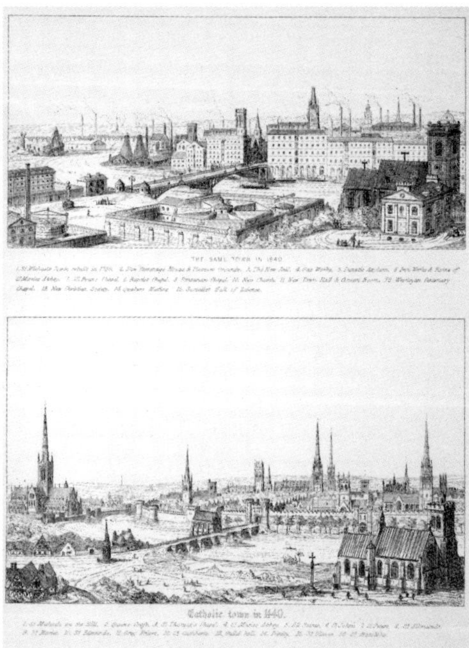

Fig. 2.4
"A Catholic town in the Middle Ages compared with how it was imagined to be in the nineteenth century," A. W. N. Pugin, 1836

Beginning in the late eighteenth century, architects discussed and visualized smoke as a form of nature that must be dealt with in industrialized societies. On the one hand, architects have attempted to curb the environmental onslaught of industrial modernization by harnessing smoke; on the other, architects have grappled with the appearance of smoke and its perceived threat to architecture and culture. The former effort may be observed in a number of eighteenth-century French practices, including those of Claude-Nicolas Ledoux and Étienne-Louis Boullée. Both architects developed stunning images of buildings releasing smoke into their surroundings. The latter type of engagement appears in a range of architectural literature and imagery emerging in early- and mid-nineteenth-century England. The industrial and smoky transformation of the towns of the English Midlands drew the attention of several architectural and spatial commentators. [**Fig. 2.2**] The imagery of English industrialization in architectural theory is particularly illuminating, especially considering its severity. England was one of the first nations to industrialize at a massive scale and one of the most innovative in developing the industrial architecture of factories. Numerous European emissaries traveled to tour the factories of England in the early nineteenth century and their impressions ranged from awe to disgust.

The 1826 English diaries of German architect Karl Friedrich Schinkel partially illustrate how architectural theories of smoke were born in the English Midlands. Schinkel's journey to the region has been noted by architectural historians for its particular focus on industrialized construction methods, but less examined is his focus on the evidence of smoke in the industrialized towns he toured. [**Fig. 2.3**] In his diary, he wrote of the "smoke from hundreds of tall obelisks," the frightening sight of a "grey, smoke-filled town built on hills and in valleys" and towns where "buildings appear as blackened with smoke as if they had been in use for a hundred years. It makes a dreadful and dismal impression."[1] In a letter to his wife, he remarked on the impropriety of the smokestacks, a trope that will continue throughout modern architectural theory engaged with the problems of industrial buildings. He wrote of "thousands of smoking obelisks from steam-engines... whose height of eighty to one hundred feet destroys all the impact of the church towers."[2] This image of the smokestack-laden English Midlands also appears in architectural discourse in the book *Contrasts* by English architect A. W. N. Pugin. In this contemporaneous work, Pugin visualized the Midlands factory town as a moral outrage and used its image to enormous rhetorical effect. Like Schinkel, Pugin comments on the relation between smokestack and church spire, but for him the smoke-dominated factory town is both an abomination of propriety and a symbol of a largely secular political economy. Pugin, a devout Catholic, viewed the smoke-ridden landscapes of the Midlands as indicative of a modern capitalist society without core religious beliefs. Pugin's interpretation of the industrial town contrasts with an earlier fifteenth-century image of the same town dominated by Gothic church spires versus industrial smokestacks. [**Fig. 2.4**] The noticeably less smoky and far more picturesque medieval town illustrates the harmony of precapitalist religious society.[3]

Pugin's outrage at the industrial town and his moral turn toward medievalism would eventually be married with the German idealistic philosophy that fueled Schinkel's views in what later became known as the English Arts and Crafts movement. Within this movement, industrialization—its architecture, exhalations, and working conditions—was viewed as a total environment that overdetermines society and culture as a whole. Such thinking was exemplified by one of the movement's chief conceptualizers, John Ruskin. Like Pugin, Ruskin wrote of the English Midlands as smoky, spoiled, and denatured wastelands. But his own theory of smoke was more complex; within the context of industrialization taking over the English countryside, he increasingly romanticized the seemingly benign smoke of the humble country home. Smoke is imagined as the center of a moral industry with the fireplace being the seat of domestic work. For example, in *The Poetry of Architecture*, Ruskin wrote of the chimneys emerging from the residences of the English countryside in Romantic terms, strongly contrasting with those of more wealthy dwellings. A vociferous critic of industrial smoke, he wrote:

> There is no motion more uniform, silent or beautiful than that of smoke; and, therefore, when we wish the peace or stillness of a scene to be impressive, it is highly useful to draw the attention to it. In the cottage, therefore, a building peculiarly adapted for scenes of peace, the chimney, as conducting the eye to what is agreeable, may be considered as important, and, if well managed, a beautiful accompaniment.

Such humble smoke is contrasted to that of laborers within larger domestic buildings. Ruskin states:

> But in buildings of a higher class, smoke ceases to be interesting... the associations it arouses are not dignified; we may think of a comfortable fireside, perhaps, but are quite as likely to dream of kitchens, and spits, and shoulders of mutton.... Now, when smoke is objectionable,

Fig. 2.5
The smoke-filled air of Chicago, Illinois, 1920s

> it is certainly improper to direct attention to the chimney... where smoke is ugly, decoration directs attention to its ugliness.[4]

Ultimately, we may discern that Ruskin's theory of smoke is focused on issues of class, labor, and the particular economic system that produces it. Smoke is not inherently vile; it is vile when its odors and colors evoke the unfair economic systems that were transforming English life; thus, it is desirable when it strongly contrasts with the smoke of industry and imparts the atmosphere of an earlier, seemingly more moral economy.

By the late nineteenth century the modern industrial city was fundamentally understood as a smoke-laden entity. Manchester, Birmingham, and London in England; Bremen, Berlin, and Essen in Germany; and Pittsburgh and Chicago in the United States were considered some of the smokiest cities of the industrialized world. [**Fig. 2.5**] Within modernity, smoke is both matter and metaphor for the emerging paradoxes of modern urbanization, and this extends from industrial to domestic smoke as well as to the act of smoking tobacco and to the imagined smoke from human bodies. As the contemporary historian Alain Corbin said, within a society stratified by class relations, by the very factory system examined by Schinkel, Pugin, and Ruskin, smoke is a signifier of class, and the ability to command one's smoky exhalations is a signifier of class position. What type of smoke one engages with and the kinds of smoke that cling to us are indicative of social rank and the level of command one has over his or her environment. Coal, wood, and tobacco smoke become new signifiers in the city.[5]

Fig. 2.6
Soot stains on the Reliance Building, Chicago, Illinois, circa 1910

Fig. 2.7
A portrait of Ludwig Mies van der Rohe smoking, *Life* magazine, 1957

Within the city of Chicago, in particular, various negotiations of these types of smoke became the practice of a new modern architecture and a new type of architect. Stretching from designs for skyscrapers, to residences, to the ways in which architects were portrayed in photographs, Chicago emerged as another key site of architectural engagement with smoke. One of the first buildings to purposely negotiate the dirty smoke of the city was the Reliance Building, an office building in the downtown financial district. The building's unusual cladding of non-load-bearing terra cotta has been lauded for its lightness, but it also protected the facade's exterior surfaces from the surrounding environment of coal smoke. In 1895, Charles Jenkins, a journalist with *Architectural Record* magazine, noted that with terra cotta, one was "able to wash your building and have it as fresh and clean as the day it was put up."[6] In early photographs of the Reliance Building, the structure appears as a gleaming white tower in a sooty city, but as a material object resistant to Chicago's smoky climate, the terra cotta presented contradictions. The material may have been vitreous, but it presented a flimsy skin of glass and clay, placing interiors in a more open relationship to the "vile soup" of the city. In photographs of the building from the early twentieth century, the Reliance Building's terra cotta skin appears completely overwhelmed by the surrounding urban smoke—blackened by the city's atmosphere. [**Fig. 2.6**]

In the suburbs of Chicago, Frank Lloyd Wright's exploration of the hearth and chimney as central elements in the home might be understood as offering an ironic focus within an otherwise intolerably smoky city. Wright described how fire from the hearth illuminates the family, but the smoke, as an exterior symbol of the gathered family, also imparted itself into the larger urban-industrial atmosphere that he disdained. Wright's focus on the hearth as the center of the home was influenced by concepts from Semper and Ruskin. Like Ruskin, Wright spoke out against the smoky atmosphere of the city, though he also concentrated on the fireplace as a central aspect of modern domesticity, particularly at a time when wealthy American homes increasingly utilized more advanced heating systems such as steam. It is within the very flatness and inarticulate form of Wright's symbolic chimneys that we see ambivalence.

In considering significant architectural engagements with smoke in this city, we should also consider the smoke moving through the interiors of buildings; here, as in other cities, the smoking of tobacco radically transformed the interior atmospheres of built spaces and perceptions of the city.[7] For example, the Chicago architect Ludwig Mies van der Rohe cultivated his image as a smoker of cigars to enormous rhetorical effect, often appearing with cigar in hand or amidst a swirl of smoke. In depicting Mies, photographers from *Life* magazine made an extraordinary but simple collage that linked smoking, interior space, architecture, and the city. [**Fig. 2.7**] Reclining in a chair, we see Mies releasing a plume of smoke that mingles with the plume of emanations from an industrial smokestack reflected in the exterior glass of his Lakeshore Drive Apartments. In this image, the trans-somatic link of smoking, interior, body, city, and the smoothness of modern cladding collapse together.[8] The image continues a visual trope in which smokers are depicted lost in thought, surrounded by their own smoky emissions, but the trope is adjusted to suggest the creation of new spaces through smoke—the architect's creations launch a new atmosphere into the city. Smoke is carried into the air, where it interacts with the surrounding glass architecture, once again resistant but visually open to the industrial atmosphere of the city.

The architectural visualizations of smoke formed in Chicago and the English Midlands have practically vanished over the past thirty years. Smoke is less explored within architectural thought. This coincides with efforts by Western municipal governments to eliminate all types of urban smoke. Within the last twenty years, numerous cities effectively banned factory smoke, incinerators, chimney smoke, grill smoke, and perhaps most controversially, tobacco smoke. If a city's economic health was once represented by its smokiness; now, the sign of a thriving Western city might be the complete absence of smoke. In fact, one might argue that the truly global character of a city is marked by a lack of both industrial and cigarette smoke. Responding to recent efforts to eliminate such emissions from cities, architectural critic Herbert Muschamp attempted a recovery of smoke as an architectural concept. Muschamp lamented the loss of the smoky interior atmospheres of New York City following the ban on smoking tobacco in restaurants. He argued that, in the city, smoke functions as a type of veil for obscuring the city's rough edges, adding glamour to its interiors and surroundings. He linked the elimination of smoke from the city to the larger culture of eviction that was overtaking the city, dispersing working-class sites of labor (factories) and relegating working- and middle-class residents to the city's peripheries, like smoke itself. Architects, he argued, give form to this smoke-free, cleansed urban atmosphere through a type of complacency with its underlying purge of unpleasantry.[9] Perhaps we experience this eviction of the smoker most literally in those specially demarcated smoking rooms in large urban airports, where smokers are collected into glass rooms and their enjoyment of tobacco becomes a suspect form of pleasure, a strange pause in spaces that emphasize movement, above all else. Recent architectural efforts to redesign these spaces have simply instantiated them without questioning their inherent logic.

As difficult as it is to imagine architects entangling spaces with smoke in a more direct manner, we might see a new type of engagement with smoke in some contemporary projects. Some architects increasingly train their design acumen on a more subtle development of airport and urban smoking rooms, and other architects more forcefully and convincingly speculate that due to the very banishment of smoke we might be able to experience it as both a form of unique socionature in the city and as an aspect of history. In other words, smoke is becoming a curiosity, even as it is recognized as a nuisance in many de-industrialized cities. For example, in a project to renovate and preserve Philip Johnson's Glass House, the experimental preservationist Jorge Otero-Pailos proposed reintroducing the smell of cigarette smoke into the space. Examining historical photos, similar in many ways to those of Mies and his cohorts smoking in his building designs, Otero-Pailos argued that tobacco smoke was a significant aspect of the perception of this space. The smell of tobacco smoke (as well as wood and cologne) was missing in the historical preservation of the space. In a publication on the project, Otero-Pailos developed scented panels so that readers could familiarize themselves with the scents of the original house. [**Fig. 2.8**] In a similar effort, I proposed a historical reconstruction of the noxious plumes inside a former San Francisco bus shed originally designed by SOM (1951)

Fig. 2.8
Proposal to reconstruct the smoky odor of the Glass House, Olfactory Reconstruction, by Jorge Otero-Pailos, 2008

Fig. 2.9
View of a reconstructed plume inside a former bus shed at the California College of the Arts, by David Gissen, San Francisco, California, 2008

but now used as classrooms and studios for the California College of the Arts. The project, called PlumeIdling, reconstructs the space's former exhaust clouds by projecting a videotape of smoke emanating from one of the few remaining industrial facilities in the neighborhood. These contemporary plumes are projected horizontally on the ground, making it appear more like the exhaust whorls formerly inside this shed. [Fig. 2.9] By linking the old bus shed with the current industrial plant, the project comments on how gentrification processes and the elimination of subnature work hand in hand.

At a much larger scale, the Pilot Plant by NL Architects for the industrial sector of Rotterdam suggests that a more significant reconsideration of industrial smoke is underway. In this enormous unrealized project, NL Architects proposes a type of space that brings inhabitants into a direct but protected relationship with industrial smoke, through innovative architectural forms and programming. The Pilot Plant's enormous concrete shells wrap a small, low-risk industrial facility near the port of Rotterdam, mitigating the plant's toxic emissions and enabling a programmatic rapprochement between industry, recreation, and institutional sectors. Crowning these shells is a two-story facility that contains several community programs, including a cafe, convention facilities, and outdoor recreational areas. The architects envision trees and greenery surrounding

Fig. 2.10
Engaging with smoke in the late-modern city, Pilot Plant, by NL Architects, Hoogvliet, Rotterdam, the Netherlands, 2002

Fig. 2.11
Section diagram of Pilot Plant, by NL Architects, Hoogvliet, Rotterdam, the Netherlands, 2002

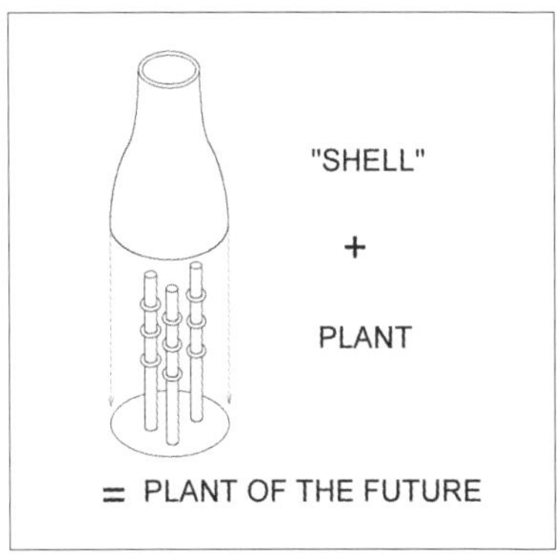

Fig. 2.12
Diagram of the Pilot Plant, by NL Architects, Hoogvliet, Rotterdam, the Netherlands, 2002

Fig. 2.13
Model of the Pilot Plant, by NL Architects, Hoogvliet, Rotterdam, the Netherlands, 2002

Fig. 2.14
View from the lecture hall of Pilot Plant, by NL Architects, Hoogvliet, Rotterdam, the Netherlands, 2002

Fig. 2.15
Aerial view of the Pilot Plant, by NL Architects, Hoogvliet, Rotterdam, the Netherlands, 2002

the concrete shells; inside the facility, views of a surrounding industrial world provide a new vision of the city. The project suggests ways that nonindustrial and industrial uses might coalesce, and the architects address pollution from industrialization as a key contemporary concern for bringing these spheres back into a physical and, in turn, social dialogue. It advances something quite different than reclaiming former industrial areas of cities by turning them into recreational sites for members of the postindustrial urban economy. It returns us to smoke and to the issues of industry, labor, and environmental degradation that smoke suggests. The Pilot Plant implies that smoke, like trees and plants, is one of many forms of nature within a city. One can only hope that this re-exploration of smoke will extend to other spheres, where smoke now appears as a peripheral and denigrated aspect of society and its practices. [**Figs. 2.10–2.15**]

Fig. 2.15
Aerial view of the Pilot Plant, by NL Architects, Hoogvliet, Rotterdam, the Netherlands, 2002

Gas

The word *gas* was coined by the scientist Jan Baptista van Helmont to describe the vaporous matter that lifted off coal when it was heated. The term hails from the Greek word for "void," emphasizing gas's inherent mutability. A vaporous substance, gas appears in architecture in factories, as fuel for artificial illumination, and from the emanations of people aggregated in cities. While we tend to associate the engineering of water systems with the modernization of cities, the provision for and negotiation of gas also played a role in urban industrialization. Architectural historian and theorist Siegfried Giedion argued that controlling gas was central to emerging concepts of modernity in the mid-nineteenth century.[1] He wrote that the production of coal gases, the use of gas for the manufacture of industrial materials, the mechanization of breadmaking with pumped gas, and the emergence of hot-air gas balloons were all interconnected attempts to harness the ineffable in industrial cities. These practices emerged at a time when scientists and inventors were fascinated by gas's ability to transform matter and space. To Giedion's list, we can add urbanists' attempts to harness and remediate the unpleasant gases associated with human and animal waste from cities. Images of the mapping, capture, and release of various gases move through some of the most exhilarating and horrifying technological concepts and experiments. Gas might be used to illuminate cities or to create landscapes of death, and these simultaneously utopian and disturbing aspects of gas are harnessed in dialectical form in contemporary architectural projects.

Some of the first architectural and urban considerations of gas involved explorations of urban miasma. The term chiefly described the gaseous odor emanating from human waste; the strong odors were feared as carriers of disease and pestilence. As was the case with the perception of smoke, the ability to perceive miasma was somewhat historically conditioned; these gases did not become sources of fear until the eighteenth century. A new sensitivity to gases coincided with a produced alienation from human waste, through transformations in urban spaces such as the separation of living and working spaces and the distancing of stables and cesspools away from everyday life within European cities.[2] Miasmatic odors were difficult to trace and to quantify, as the gases that carried them were fleeting and ephemeral. In an effort to understand their reach, nineteenth-century urban reformers began conducting olfactory surveys of urban spaces. In one of the most famous of these, English reformer Hector Gavin developed an entire cartography of

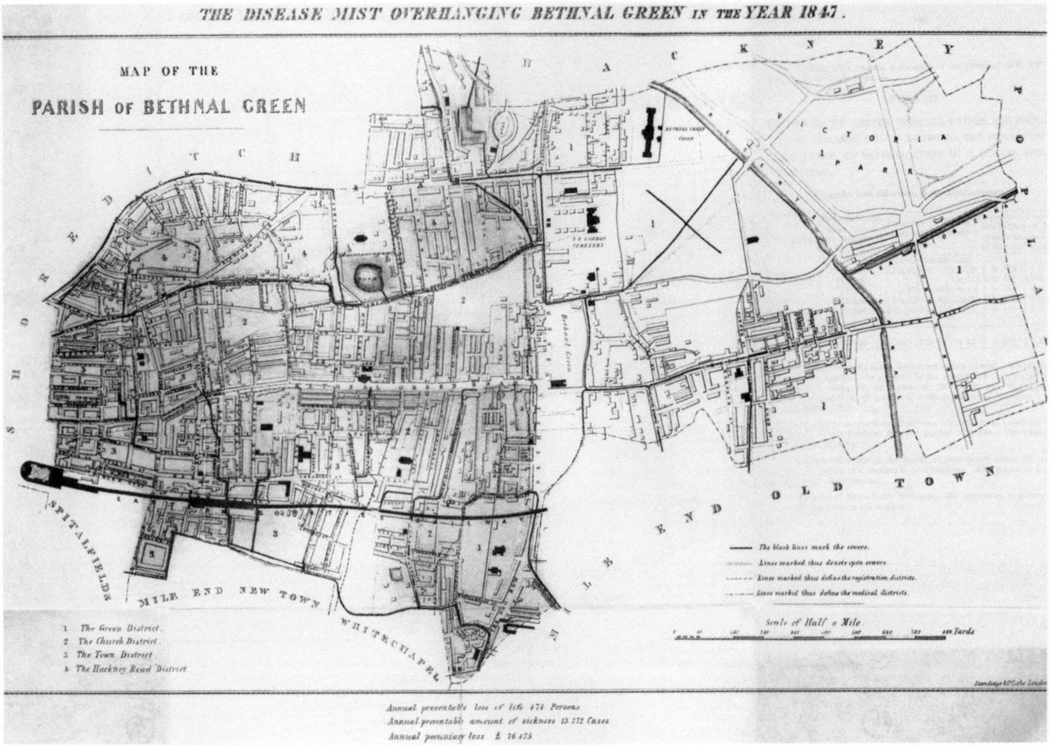

Fig. 3.1
Map of the "disease mist," Hector Gavin, 1848

miasmatic gas within the London neighborhood of Bethnal Green. [Fig. 3.1] Gavin drew a hovering cloud of stagnant brown air layered over a diagram of this working-class neighborhood's spaces and feeble sewerage system. The "pestilent disease mist," as Gavin labeled the cloud of gas, reinforced existing relations between poverty and environmental degradation.[3] With their language of objectivity, his mappings mobilized an emerging upper- and middle-urban class to simultaneously fear and address the gaseous atmosphere of modern urbanization through replanning. In the case of Bethnal Green, the neighborhood was slated for demolition and reworking to make the interiors of buildings and public spaces incapable of holding stagnant air. The solutions to ridding urban spaces of miasma also involved an increased use of plumbing systems. In London, extensive sewerage systems rid urban dwellings of potentially dangerous human and animal wastes while helping alleviate the powerful odors that overcame these neighborhoods. But as beneficial as these systems were, they only produced new forms of gases of their own. Sewer gas, which collected from the decomposing waste, became a new type of urban gas that required negotiation in urban spaces and streets. Sewer pipes filled with gases often tore apart city streets in violent explosions.

Fig. 3.2
"The Gas Works at Courcelles," Ernest Jean Delahaye, 1884

Fig. 3.3
The poison gas clouds of World War I

As neighborhoods rid themselves of undesirable gases, they were also pumped with new usable gases oriented toward other radical transformations of urban space. Coal gas, one of the chief combustible fuels of the nineteenth century, was harnessed for artificial light throughout virtually all major Western cities. Coal gas was manufactured in enormous gas factories or *retorts*, where workers raked mounds of coal into ovens; the gas given off by these smoldering chunks was collected in pipes overhead, fed into large gasometers, and then distributed to urban neighborhoods. [**Fig. 3.2**] The entire enterprise of coal gas distribution created new urban scenes in Berlin, London, Paris, and New York that "turned night into day," but coal gas factories also imparted new stenches into the city, horribly polluting their urban surroundings. One urban critic wrote of the coal gas factories:

> Wherever a gas-factory—and there are many such—is situated within the metropolis, there is established a centre whence radiates a whole neighborhood of squalor, poverty and disease. No improvement can ever reach that infected neighborhood—no new streets, no improved dwellings, not even a garden is possible within a circle of at least a quarter of a mile in diameter, and not so much as a geranium can flourish in a window-sill.[4]

Like miasmatic gas, nineteenth-century writers understood coal gas to pollute both the physical and social structures of society. It is not only abysmal, it is a form of human-produced nature that is imagined to impart both new forms of environmental and social degradation. By the twentieth century, this aspect of the production of gases became more deadly and more literal with the industrial production of dyes, the chemical vapors of which were lethal. These industrially produced gases contained concentrated chemicals as vehicles for transferring color, and they crippled workers with their toxicity. The German dye-making factories also produced poison gases for World War I, as they were equipped to handle the chief components of poison gas, chlorine and urea. If the manufacture of coal gas produced horrific working conditions, the manufacture of known lethal gases lent industrialization a new inhumanity.[5] [**Fig. 3.3**]

The engineering of gas out of the city and the manufacture of gases for industrial and military use illustrate the ways modern architecture and cities have engaged with and produced new forms of subnature, but in considering the historical relationship between modern society and gas we should consider one final aspect—the capture of ephemeral gases in architectural structures. We have briefly discussed the capture of gas in gasometers, but the containment of gas was also an important aspect of utopian imagery in architecture and urbanism of the mid-twentieth century. Historian Marc Dessauce traces this utopian sentiment to French experiments with balloons in the late eighteenth century. The ability to escape the surface of the earth through the harnessing of gas carried notions of earthly transcendence, and the ability to turn the new milieu of gas into a "thing" also enforced the notion of modernity as having transformative effects on space and time.[6] While the capture of gas appears in numerous architectural works, the utopian flavor of pressurized environments reached an apex in the experimental practices of the late 1960s and early 1970s. In the development of inflatables, architects explored building shapes pumped with breathable air, helium, and hydrogen gases. What is particularly interesting is the way architects saw inflatable architecture as providing refuge from other gases. For example, in the 1960s the architectural and art collective Ant Farm was inspired by images of protesting students fired upon with clouds of tear gas by local police, as well as by environmentalist discussions of air quality. Ant Farm sought to develop an inflatable refuge from an external environment plagued by emissions of both forms of social power—the police power wielded at the campus and potential forms of emergencies and warfare. Inside, an interior of pumped, safe, breathable air is contrasted with the potentially noxious air of political contention. Ant Farm begins to express how gas began to take on simultaneously chilling and absurd connotations.[7] [**Fig. 3.4**] The practices of Ant Farm were mirrored in the work of the Parisian Utopie group, who also used inflatables to politically expressive ends, and of course, the Archigram group in Britain.

Recent years may have witnessed a return to an earlier Victorian ethos in which our fear of socially produced natures, such as gas, drives forms of architectural and urban development. Fears of spatialized gases and the need for refuge from them has continued in contemporary discussions of architecture. In the 1970s and 1980s, inhabitants of office buildings increasingly became ill with "sick building syndrome," an ailment that causes headaches, dizziness, and nausea. Investigations into the causes of this syndrome revealed that among the culprits were the off-gassing materials and improper ventilation. Architects and manufacturers increasingly understood that industrially produced materials, such as those found in office furniture, tend to evaporate their chemical finishes and glues, many of which contain derivatives of oil and coal. Architectural scientists such as Michael Braungart maintained that a turn toward production practices that relied on inert materials, in addition to the

Fig. 3.4
A symbolic refuge from gas, the Clean Air Pod, by Ant Farm, University of California, Berkeley, 1970

maintenance of ventilation standards in buildings, would mitigate these gases in building interiors.[8] But our fears of contemporary gas also involve the exterior of buildings. More recently, fears of an urban gas attack have led architects and engineers to collaborate on methods for turning office buildings into refuges that protect their inhabitants from the external atmosphere. In Times Square, engineers funded by the United States Department of Homeland Security developed pressurized building technologies to work in a manner that is precisely the opposite of those envisioned in green designs. These building skins are designed to limit the exchange of air between interior and exterior. Additionally, fears of terrorism in the midtown neighborhood have led scientists to develop new ways of charting the movement of gases in cities, in many ways echoing the early work of gas cartographers such as Hector Gavin.

Within this troubling contemporary milieu, recent architectural projects explore our discomfort, dislike, and fear by conceiving of gas in new ways, different from the experimental work of thirty years ago and from recent green designs. Among contemporary works, we might look to Lars Spuybroek's Blow Out project as attempting a rapprochement with the seemingly most taboo of gases. In this design for a public restroom in an urban park, Spuybroek approaches, if not ridicules, the historically conditioned fear of human-produced gases. [**Figs. 3.5 + 3.6**] Blow Out is a design for a public toilet that both eliminates and propels human-produced gases into the surrounding space. Spuybroek writes of the project:

> This building establishes a dynamic equilibrium between internal pressure ("gotta go") and huge external forces. It is modeled in such a way that the wind blows through it at high speed (with the "grill" on the male side and the "exhaustion-pipe" on the female side). The doors in the completely distorted passage are too large for the limited space and this, together with the vector of the wind, produces, prior to total release on the luxury toilet, an additional external pressure of increased intimacy... the vector of the wind carries the smell of others, their noises, their interiors. This liquid machine connects one interior with another, it shapes intimacy, builds it up, and releases it. Finally, sitting on the toilet, orthogonal to the direction of the geometry, one can relax and let go.[9]

Mixing humor with the ridiculous, Spuybroek produces the type of territory of gas once mapped by Victorians. Gas is positioned as a breach between the imagined line that separates the individual from a surrounding space. A somewhat different but still territorial concept of gas is taken up in the project Big Leaks. In this work, Jeffrey Inaba asks us to consider accidental releases such as gas disasters, radioactive leaks, and oil spills as a type of architecture that might be controlled. Rather than just fearing and lamenting these occurrences, he takes the somewhat radical position that the leak could be envisioned as something that might be designed. [**Fig. 3.7**] Accompanying a map of these leaks and their devastating effects, Inaba writes:

> In civil, mechanical, and environmental engineering the prevention of releases of volatile, combustive, or hazardous materials often involves the design of a leak...to avoid a catastrophic big leak. When a massive failure happens, engineers are hurried in to design a response. Apart from the emergency act of containment, where the immediate spread of the seepage is impeded, most often a designed leak is implemented to divert or dissipate the material as part of the process of recovery and remediation. Instances such as Love Canal, Bhopal, and Warri plainly show that it is when these latter leaks are not introduced that harmful substances remain, causing further contamination...it is conceivable to operate at the scale of a city or regional environment using extensive means and manpower to engineer a fix in the face of disaster.[10]

Fig. 3.5
A building for moving gas, Blow Out, by Lars Spuybroek/ NOX, Neeltje Jans, the Netherlands, 1997

Fig. 3.6
Interior of Blow Out, by Lars Spuybroek/NOX, Neeltje Jans, the Netherlands, 1997

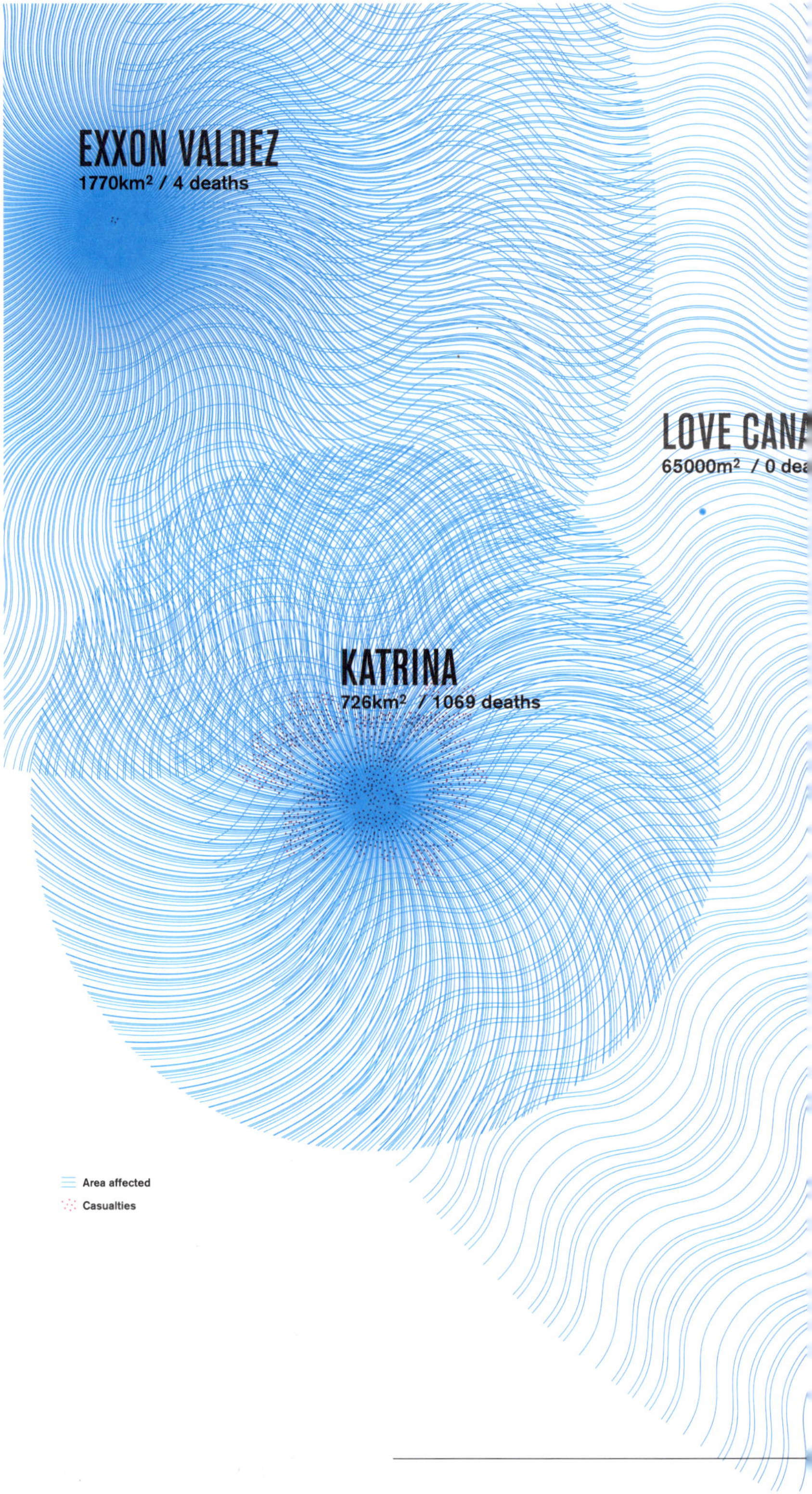

Fig. 3.7
A map of gaseous clouds and other deadly leaks, Big Leaks, by Jeffrey Inaba/C-Lab, 2005

CHERNOBYL
10600km² / 4065 deaths
BHOPAL
20km² / 20000 deaths
WARRI-KADUNA PIPELINE
7060m² / 1082 deaths

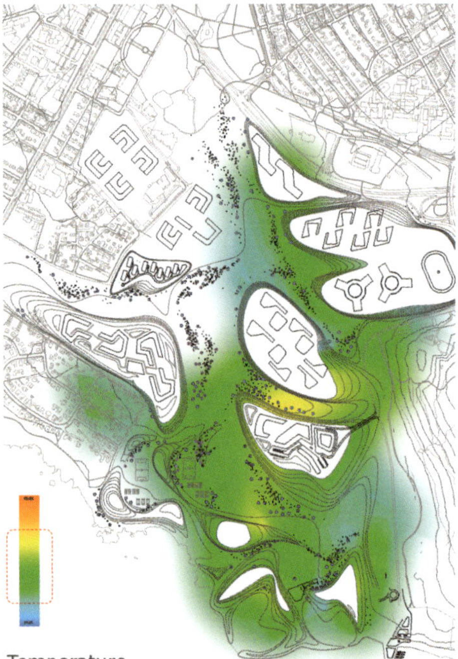

Temperature
(Increased & Controled Geo-Thermal)
- WINTER - December

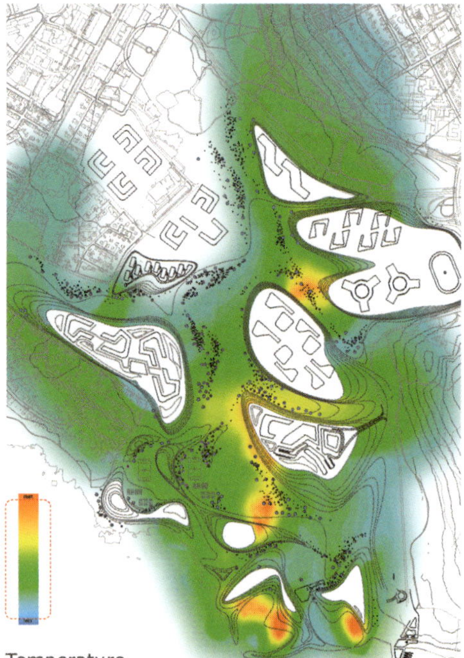

Temperature
(Increased & Controled Geo-Thermal)
- SUMMER - June

Winds
-December (east)

Winds
-June (north)

Inaba is interested in the territory of these spaces as both disasters and as engineered failures. Modern societies must confront enormous forms of risk, and in this project, concepts of risk become conjoined with a spatial program. He is not arguing in favor of poisonous matter being released into populated areas; rather, he is arguing that the idea of leaks, of the frightening moments when material transgresses boundaries, should become part of the design approach in order to avert potential disasters. Lastly, in Sean Lally's proposal for the port of Reykjavík, the design of leaks informs the entire strategy of an urban plan. He proposes tapping into the thermal energy of the earth, releasing clouds of warmed air to produce new zones of pleasurable, inhabitable space that will then generate a new form of urban planning. [**Figs. 3.8 + 3.9**]. In this project, the fearsome territories of Spuybroek and Inaba are reconsidered, and the release of gas enables a new form of social aggregation. Lally produces a vision of cities in which individuals gather together within a cloud of gaseous vapors. [**Fig. 3.10**] Perhaps this experience is only possible in contexts in which actual fears of the atmosphere are mitigated or suspended. This vision suggests that the utopian image of captured gas will be replaced by amore amorphous one.

Fig. 3.8 (opposite)
An urban plan based on the gaseous vapors of the city, Vatnsmyri Urban Plan, by Sean Lally/Weathers, Reykjavik, Iceland, 2007

Fig. 3.9 (right)
Plan detail, Vatnsmyri Urban Plan, by Sean Lally/Weathers, Reykjavik, Iceland, 2007

Fig. 3.10 (overleaf)
People enveloped in gaseous cloud, Vatnsmyri Urban Plan, by Sean Lally/Weathers, Reykjavik, Iceland, 2007

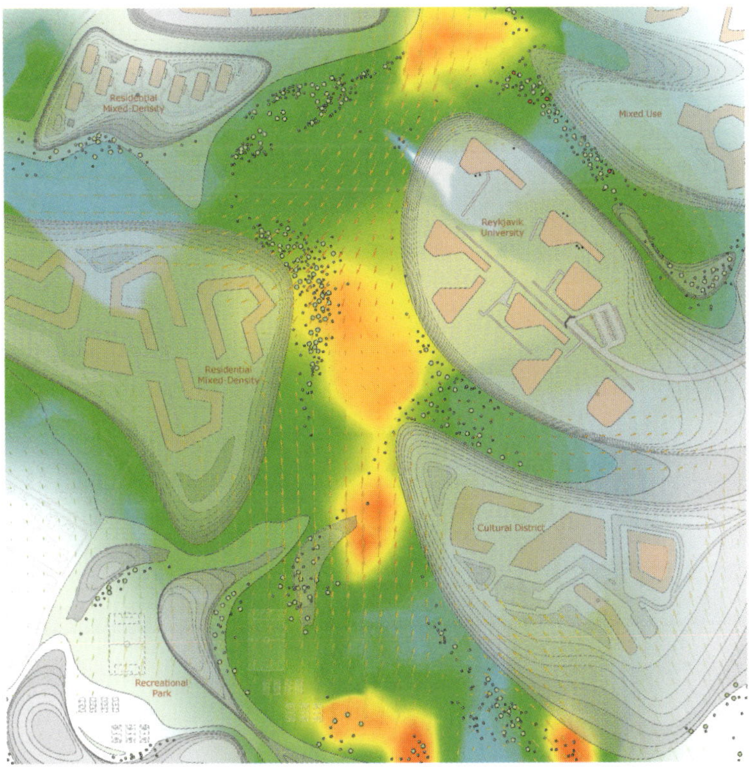

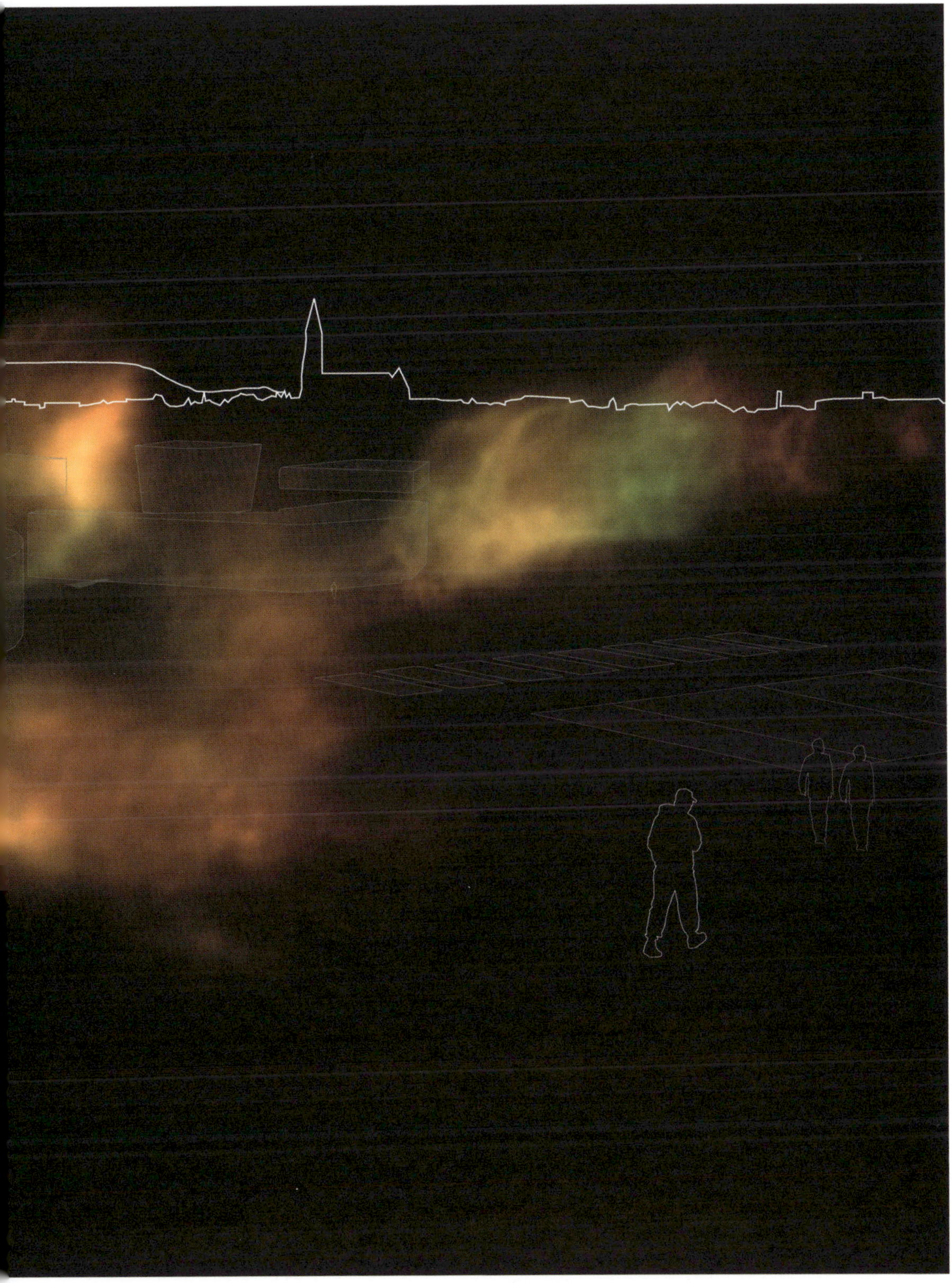

Exhaust

We have explored the origins of architecture in dankness, the formation of smoke in spaces and cities from pre- to early modernity, and the negotiation of gas in modern practices. We conclude our examination of atmospheric subnatures by examining how architects have engaged exhaust. These vaporous and malodorous emissions from automobiles are the only form of subnature explored in this book that cannot be traced to a premodern phase of architecture. As the byproduct of vehicles, roads, and highways, exhaust is largely a twentieth-century phenomenon. The experience of exhaust in cities is similar to that of smoke and gas. That is, modern cities are ridden with exhaust despite efforts by planners and urban engineers to control and harness it. The removal of exhaust has involved not only the redesigning of roadways but the creation of new forms of architecture, such as exhaust towers, and new forms of urbanism to reconceptualize the relationship between buildings and roads. In total, we might argue that the attempted elimination of this modern subnature is intimately related to the development of architectural forms of coordination and integration between buildings and roadways—what would eventually lead to the concept of the *urban system* and the monumental form known as the *megastructure*. As fantasies of urban coordination diminished, architects simply addressed exhaust through forms of disengagement. Increasingly, they sealed building interiors from the real and imagined threats of the roadway. Exhaust would continue to be examined by environmental planners and automobile technicians and would become part of the material language of those subcultures that expressed themselves within the dynamics of automobile engines. Only recently has exhaust re-emerged in startling new forms of architectural expression by experimental practices.

Some of the first architectural ideas regarding the control of vehicle exhaust appear in the work of Le Corbusier. In his reflections on Paris, Le Corbusier depicted the inadequacies of existing nineteenth-century street networks in response to environmental pressures for improved mobility. Le Corbusier did not write of exhaust as a distinct substance; rather, he considered it part of the general unpleasantness of roadways, which he referred to as "noise, gasoline stench, and dust."[1] Le Corbusier sought ways to mitigate the general effluvia that vehicles stirred up within cities; this included both their emissions and the atmosphere churned up by them when moving down streets laced with ash and dried animal manure. This involved strategies of sequestration; moving vehicles away from spaces of

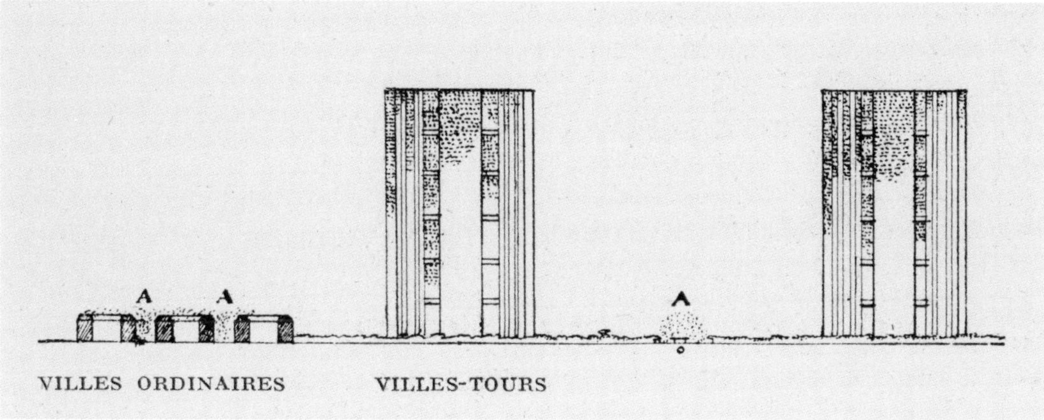

Fig. 4.1
***A City of Towers*, illustration of a city set apart from automobile pollution, from *Vers une architecture*, Le Corbusier, 1923**

habitation and into distinct routes that carried cars through cities but were surrounded by open spaces. In an important drawing from *Vers une architecture* (first translated as *Towards a New Architecture*), Le Corbusier compares the nineteenth-century city, where vehicles pumped their emissions into the built surroundings, to a new urban roadway surrounded by greenery. [**Fig. 4.1**] He believed that a buffer of greenery would provide refuge from vehicle exhaust, a concept that extends attitudes toward streets developed by nineteenth-century British reformers. Equally significant was his belief that if people were aggregated into housing towers, it would be possible to further mediate the deleterious effects of modern urbanization. For Le Corbusier, the tower promoted a better urban environment, as it permitted the planting of more green space and the surrounding of urbanites with fresh air.[2] While the *tower in the park*, as latter aspects of this strategy were eventually termed, was not conceived solely in relation to urban road pollution, this illustration demonstrates that the presence of such matter certainly informed his thinking.

Le Corbusier's ideas and those of the Parisian urbanist Eugene Henard influenced planning in New York, where the construction of new roadways and bridges became an important aspect of regional development. The movement of automobiles through the city was not just an issue of traffic circulation but also air circulation. In the proposal for the Chrystie-Forsyth Parkway, Le Corbusier's pairing of verdure-clad highways with residential apartment towers was re-envisioned for American cities. In addition to the isolation of roadways with greenery, we see how towers set away from the road play a significant role in exhaust-proofing the city. Planners in New York also considered the presence of exhaust in other circulatory constructs, such as tunnels. With the construction of the Holland Tunnel, one of the first large-scale tunnel projects explicitly devoted to automobile traffic, we witness how engineers effectively urbanized exhaust in their efforts to remove it. The Holland Tunnel spanned an enormous 8,500 feet. At each end, engineers designed ten-story ventilation towers that would push air through tunnels above the cars, drawing the vehicle exhaust upward, where it would be blown back through the tops of the towers and over industrial areas of the city. [**Fig. 4.2**] The exhaust towers provided a strange new building type in the city—a looming blank tower that oscillated between a work of engineering and architecture. Critic Lewis Mumford proclaimed that the ventilation towers of the Holland Tunnel heralded a new advance in building that collapsed architecture and engineering. Mumford wrote of these strange buildings:

> While the path toward an appropriate modern architecture was kept open by the excellent individual work of Frank Lloyd Wright, a corresponding communal advance was being made by the engineers who standardized building processes, invented new units of heating and plumbing and occasionally, almost without knowing it, threw up fine engineering structures of their own, such as the Ford Plant at Baton Rouge or the ventilator units of the Holland Tunnel in New York. Conscious of quantitative relations alone, impervious to the human effects of their processes, innocent of the aesthetic result, the engineer nevertheless had a real contribution to make. It was through his inventions and processes that architecture ceased to be the concern solely of the carpenter and the stonemason: a new battalion of trades and techniques entered it.[3]

For Mumford, the engineering of exhaust in the city as a quantitative problem of atmospheric dynamics enabled a new type of architecture that matched the novelty of the particular environmental problem it was designed to solve. The Holland Tunnel—its underground tubes and ventilators—was a new typology that was both engineered infrastructure and architecture.

The fantasies of coordination between engineering, architecture, and infrastructure expressed in the Holland Tunnel reached an apex in postwar proposals for urban precincts integrated within, around, and over roadways, such as in Paul Rudolph's renderings of the proposed Lower Manhattan Expressway. [**Fig. 4.3**] This peculiar and intense embrace of the roadway by architects was mostly due to real estate pressures in dense urban areas that made the open area above roadways attractive sites for relatively cheap development. Of the postwar proposals by architects such as Paul Rudolph, Ulrich Franzen, and others,[4] one of the more famous was the Washington Bridge Extension Complex of 1963. The entire structure, which

Fig. 4.2
The Holland Tunnel exhaust tower, 1926

Fig. 4.3
Lower Manhattan Expressway, drawing by Paul Rudolph, 1972

spanned the narrow end of Upper Manhattan, included a new highway extension topped by a bus station and four apartment towers. The complex fed and received traffic from the George Washington Bridge and the Bronx. Pier Luigi Nervi designed the bus station, and in it he developed a design in cast concrete that would purposely move roadway exhaust through and over the building. [**Fig. 4.4**] A series of triangular vents rise over the bus station's main receiving area, providing ventilation and light. For the apartment towers, architects Brown + Guenther took an opposite approach. They attempted to seal the building from the roadway's fumes, providing a space for safe habitation above the highway. For architectural historian and theorist Reyner Banham, the Washington Bridge Apartment Complex's integration of roadways, service, and apartment buildings represented a new type of architectural assemblage, the *megastructure*. Within such megastructures, architectural form, infrastructural systems, and urban planning coalesced into monumental objects that appeared to rectify the disconnections between modern architecture and urban context. In light of the automotive imagery within so many infrastructural projects, we might view the megastructure as an attempt to resolve the automobile's often enormously negative impact on

Fig. 4.4
Washington Bridge Extension Complex, view from the air, 1963

the city's environment. Megastructural projects often appear to bring mobility into the more traditional language of infrastructural systems—catchments, reservoirs, and distribution hubs—that we associate with water. Further, several megastructural projects contain robust exhaust towers rising above roadways—a possible nod to the architectural iconography of the Holland Tunnel, in which architects and engineers first sought to systematize highway exhaust.[5]

As architects proposed more tightly coordinated spaces of mobility and habitation, engineers had more carefully identified the specific chemical composition of exhaust and its potential health effects. Exhaust was composed of a number of dangerous substances, including aerosolized lead, carbon monoxide, and hydrocarbons. In certain concentrations, these components of gas produced disorientation, physical pain, and potentially death. Historians of the mid-twentieth century have noted Western fears of cities consumed in a holocaust of radioactivity, but a review of newspapers from the same period reveals that many individuals also feared that the city would be consumed by clouds of automobile exhaust.[6] The impending ruin of the modern city was captured in photographs of exhaust clouds as much as in imagined billows of radioactive fallout. [**Fig. 4.5**] Although such whorls were primarily tied to industrialization, in the 1950s and 1960s *exhaust* had slippery associations. It was associated with certain subcultural groups that imagined roadways as sites of expression. With the rise and eventual downturn of coordinated urbanization, exhaust became the dominant subnature of the road and the matter expelled by motorcycle gangs and drag racers, both of which viewed roads as sites of personal transformation, signifying the power of mobility.

Due to both the inherent dangers of linking urban spaces with exhaust-ridden roadways and the enormous costs associated with such projects, by the mid-1970s, the idea of using urban planning to either utilize the area above highways or to alleviate automobile-related pollution begins to appear increasingly naive. For example, by the mid-1970s, the Washington Bridge Apartments were deemed a failure, as the buildings became inundated with exhaust and the bottom four floors were virtually uninhabitable.[7] Rather than attempt a massive recoordination between buildings and roadways, a more reasonable approach to remediating exhaust involved developing new legislation for lowering the emissions of automobiles and protecting individual buildings by removing them from direct engagement with highways, as well as through improvements in building-skin technology. The use of catalytic converters and other technical means in

Fig. 4.5 Observing the smog-filled air of New York City, circa 1953

automobile engines helped control pollution in Western cities, and instead of establishing new planning standards, architects sought ways to restructure individual buildings to protect inhabitants against exhaust, particularly for intensely urban building types such as skyscrapers. For example, the development of double-skin facades in skyscraper construction helps filter outdoor pollutants even though their primary use is energy efficiency.

Like smoke, exhaust is being expunged from daily experience in the city in the name of environmental health, but in a manner that suggests unarticulated forms of social control. Exhaust is tied to particular vehicles and spaces that also invoke specific practices and urban classes—buses, bus stations, motorcycles, older vehicles, and raceways. Exhaust is also linked to particular geographies, as developing countries, such as China and India, are often deemed incapable of controlling vehicle exhaust. Such assessments position the Western urban environment as superior to that of developing countries. In this sense, exhaust represents a denigrated and less sophisticated urban environment from an environmental, social, and geographical point of view; in many ways, exhaust represents an uncivilized past from which a more clean and global city will emerge.

It is within this context that we might consider a recent project that explores exhaust as an environmental, urban, and social form—the B_mu Tower by the architectural firm R&Sie(n). In the B_mu Tower (2002), R&Sie(n) suggests more provocative ways to conceive architecture's engagement with the dusty, dirty exhaust-ridden air of urban roadways. This project is less about providing protection from exhaust or eliminating such nuisances altogether, and more about enabling us to understand our historically conditioned attitudes toward urban pollution and the ironic posture we often take when attempting to remediate it. For the B_mu Tower, R&Sie(n) was asked to design a highrise—containing a gallery, offices, and cafe—for the polluted context of downtown Bangkok. [**Fig. 4.6**] Rather than simply create a climate-controlled and environmentally filtered space for the people of the city, the firm used the building to "breed" the exhaust of the city onto the structure's exterior surfaces; they went as far as developing an inspirational image of a woman being enveloped by automotive exhaust. [**Fig. 4.7**] The firm organized the building into several architectural volumes that maintained the contemporary parameters of internal environments; surrounding this multilevel stack, they wrapped these spaces in an electrostatic skin that actually draws the exhaust of the street toward the structure. R&Sie(n) ultimately developed a new interior for the pollution-ridden environment of Bangkok that considers the role of the tower as refuge from a spoiled environment—a role we have seen from the early twentieth century to the present. [**Fig. 4.8**] But as aggressive as this project is, the skin that attracts the dusty exhaust also filters air, partially on the outside and more thoroughly on the inside, in order to maintain standards of health and conservation inside the space. [**Fig. 4.9–4.11**] In this way, R&Sie(n) designed a building that simultaneously pulls exhaust toward it while developing a protected context from it. It enables us to view the pollution of the city within a programmatic context and formal type known for a high degree of environmental control—an art gallery and a tower. This project also enables us to see the ironies and potential loss of historical understanding inherent in seeking to achieve an ever-more depolluted and rarified milieu. In bringing us close to exhaust, R&Sie(n) enables us to see its expulsive vapors, striking textures, climatic effects, and qualities. [**Figs. 4.12 + 4.13**]

Fig. 4.6
The exhaust-filled air of Bangkok, Thailand, circa 2000

Fig. 4.7
Conceptual study for the B_mu Tower, by R&Sie(n) Architects, Bangkok, Thailand, 2002

Fig. 4.8
Drawing of the B_mu Tower emerging from vaporous air, by R&Sie(n) Architects, Bangkok, Thailand, 2002

Fig. 4.9
Perspective rendering of the B_mu Tower, by R&Sie(n) Architects, Bangkok, Thailand, 2002

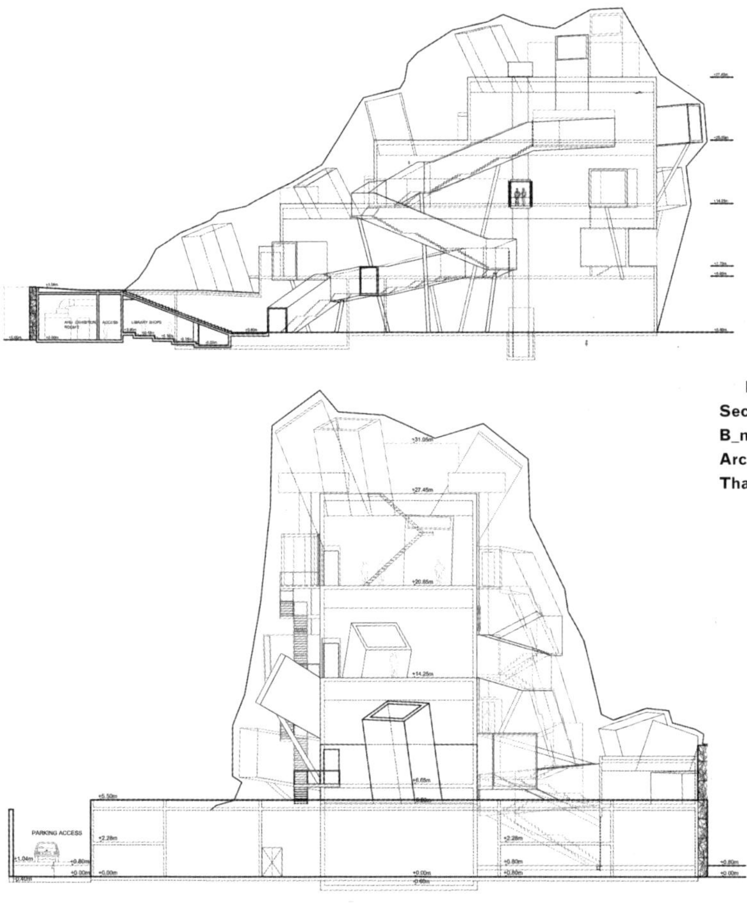

Fig. 4.10
Section diagrams of the B_mu Tower, by R&Sie(n) Architects, Bangkok, Thailand, 2002

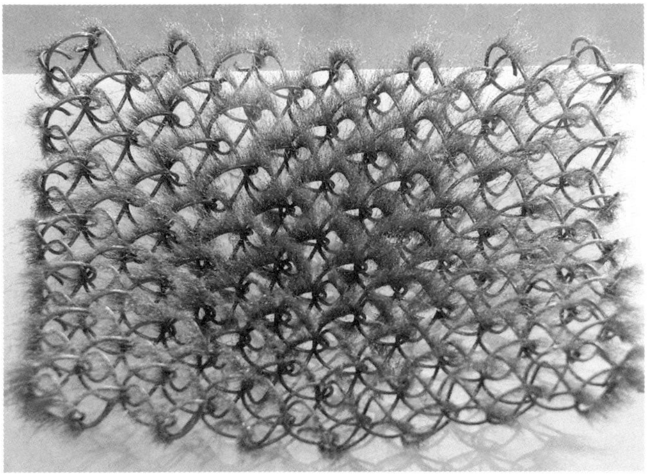

Fig. 4.11
Skin detail, B_mu Tower, by R&Sie(n) Architects, Bangkok, Thailand, 2002

Fig. 4.12
Conceptual section diagram, B_mu Tower, by R&Sie(n) Architects, Bangkok, Thailand, 2002

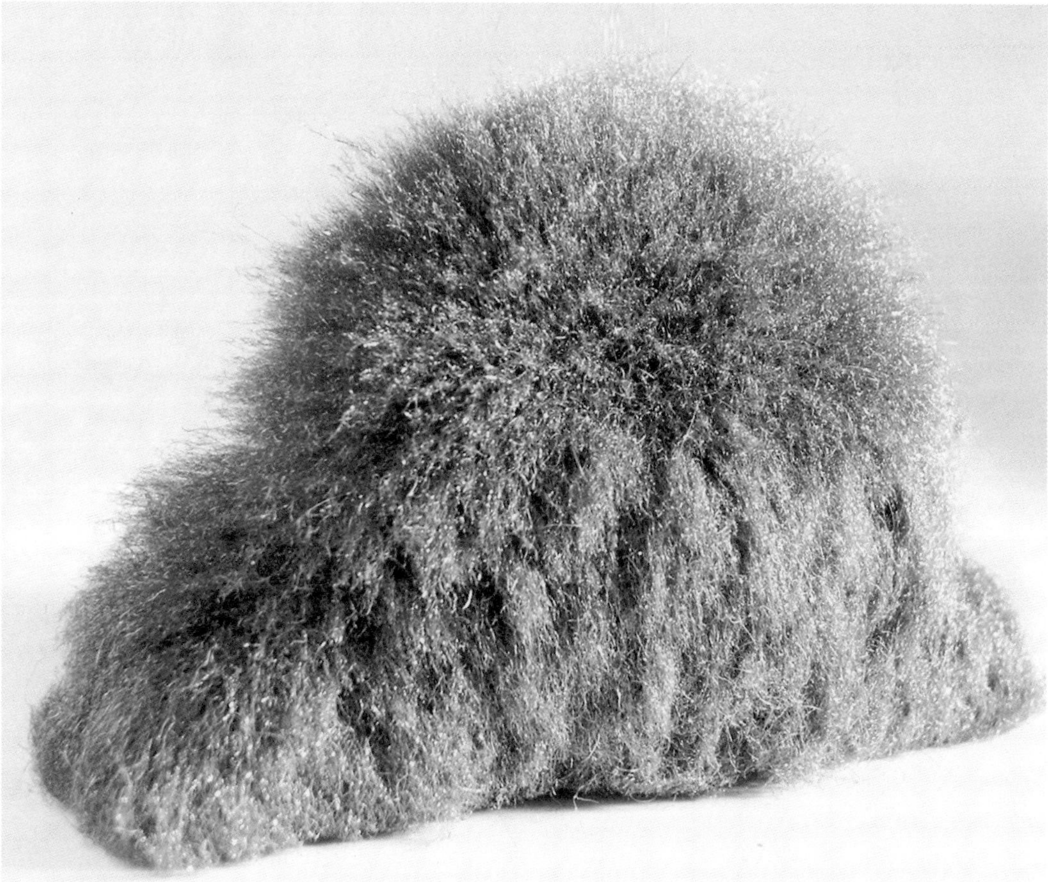

Fig. 4.13
Conceptual study model, B_mu Tower, by R&Sie(n) Architects, Bangkok, Thailand, 2002

Dankness

1. Marc-Antoine Laugier, *An Essay on Architecture*, trans. Wolfgang and Anni Herrmann (Los Angeles: Hennessey & Ingalls, 1977), 11.

2. For further examinations of this literature see Joseph Rykwert, *On Adam's House in Paradise: The Idea of the Primitive Hut in Architectural History* (New York: Museum of Modern Art, 1972); and Harry Francis Mallgrave, *Modern Architectural Theory: A Historical Survey*, 1673-1968 (Cambridge, UK: Cambridge University Press, 2005).

3. On the development of the grotto see Naomi Miller, *Heavenly Caves: Reflections on the Garden Grotto* (New York: George Braziller, 1982).

4. See Victor Hugo, "The Intestine of the Leviathan" in *Les Misérables* (New York: Signet, 1987), 256-75. Matthew Gandy, "The Paris Sewers and the Rationalization of Urban Space," *Transactions of the Institute of British Geographers* 24, no. 1 (1999): 23-44.

5. Le Corbusier, *Precisions: On the Present State of Architecture and City Planning*, trans. Edith Schreiber Aujame (Cambridge, MA: MIT Press, 1991), 38.

6. Ibid., 38-40.

7. Gaston Bachelard, *The Poetics of Space*, trans. Maria Jolas (Boston: Beacon, 1994), 23.

8. See Anthony Vidler, *The Architectural Uncanny: Essays in the Modern Unhomely* (Cambridge, MA: MIT Press, 1992), 118-45.

9. Philippe Rahm, *Underground House*, board from the exhibition Anxious Climate, 2007.

10. See Grotto in Benjamin Aranda, *Pamphlet Architecture 27: Tooling* (New York: Princeton Architectural Press, 2006), 80-89.

Smoke

1. Karl Friedrich Schinkel, *The English Journey: Journal of a Visit to France and Britain in 1826*, ed. David Bindman and Gottfried Riemann (New Haven: Yale University Press, 1993), 140.

2. Ibid., 180.

3. A. W. N. Pugin, *Contrasts: Or, A Parallel Between the Noble Edifices of the Middle Ages and Corresponding Buildings of the Present Day, Shewing the Present Decay of Taste* (1836; Edinburgh: John Grant Publishers, 1898).

4. John Ruskin, *The Complete Works of John Ruskin*, vol. I (New York and Chicago: National Library Association, 1905), 62-63.

5. Alain Corbin, *The Foul and the Fragrant: Odor and the French Social Imagination* (Cambridge, MA: Harvard University Press, 1986).

6. Quoted in Joanna Merwood, "The Mechanization of Cladding: The Reliance Building and Narratives of Modern Architecture," *Grey Room* no. 4 (2001): 52-69. See this essay for more on the image of smoke in the Reliance Building.

7. Some of the thinkers we have been exploring throughout this study of smoke carried strong opinions about tobacco smoke's ultimate effects. Ruskin, the cartographer of smoke, was virulently antismoking, being parodied in an antismoking ad by the British tobacco company, Copes. His disciple, William Morris, had great appreciation for smoking, and was lauded as one of the many Victorian intellectuals who "worshipped at the altar of Nicotina."

8. For more on this image see the essay by Beatriz Colomina, "Mies Not," in *The Presence of Mies*, ed. Detlef Mertins (New York: Princeton Architectural Press, 1994), 193-202.

9. Herbert Muschamp, "Thank You for Not Smoking," *New York Times*, May 14, 1995.

Gas

1. Siegfried Giedion, *Mechanization Takes Command: A Contribution to Anonymous History* (New York: Oxford University Press, 1948), 180-87.

2. Alain Corbin, *The Foul and the Fragrant: Odor and the French Social Imagination*, trans. Miriam Kochan (Cambridge, MA: Harvard University Press, 1986).

3. On Gavin's map and the charting of miasma, see "Rookeries and Model Dwellings: English Housing Reform and the Moralities of Private Space," in Robin Evans, *Translations from Drawing to Building and Other Essays* (Cambridge, MA: MIT Press, 1997); and Erin O'Connor, *Raw Material: Producing Pathology in Victorian Culture* (Durham, NC: Duke University Press, 2000).

4. Lynda Nead, *Victorian Babylon: People, Streets, and Images in Nineteenth-Century London* (New Haven, CT: Yale University Press, 2000), 94. On gasworks, from a more theoretical perspective, see Steven Connor, "Gasworks," *19: Interdisciplinary Studies in the Long 19th Century*

5. See Steven Connor, "An Air that Kills: A Familiar History of Poison Gas," http://www.bbk.ac.uk/english/skc/gas.

6. Marc Dessauce, *The Inflatable Moment: Pneumatics and Protest in '68* (New York: Princeton Architectural Press, 1999).

7. See Felicity Scott, *Architecture or Techno-Utopia: Politics after Modernism* (Cambridge, MA: MIT Press, 2007), 208-45.

8. See the various notes on off-gassing and material responses in David Gissen, ed., *Big and Green: Toward Sustainable Architecture in the 21st Century* (New York: Princeton Architectural Press, 2003).

9. See http://www.archilab.org/public/1999/artistes/noxa01en.htm.

10. Jeffrey Inaba, "Big Leaks," *Volume* 5, 2005.

Exhaust

1. Le Corbusier, *Towards a New Architecture* (Los Angeles: Getty Publications, 2007), 125.

2. Ibid., 124.

3. Lewis Mumford, *The Brown Decades: A Study of the Arts in America, 1865-1895* (New York: Dover, 1955), 81.

4. For some of these schemes, see Peter Wolf, *The Evolving City: Urban Design Proposals by Ulrich Franzen and Paul Rudolph* (New York: Whitney Library of Design, 1974).

5. Reyner Banham, *Megastructure: Urban Futures of the Recent Past* (London: Thames and Hudson, 1976), 30-32.

6. Robert A. M. Stern, Thomas Mellins, and David Fishman, *New York 1960: Architecture and Urbanism Between the Second World War and the Bicentennial* (New York: Monacelli Press, 1995), 99-100.

7. For the history of these buildings see David Gissen, "Exhaust and Territorialisation at the Washington Bridge Apartments, 1963-1973," *Journal of Architecture* 12, no. 4 (2007): 449-61.

MATTER

Dust

We now switch from focusing on atmospheres to matter. Perhaps no form of subnatural matter haunts architecture more than dust. It appears in views of Roman ruins by Giovanni Battista Piranesi, in drawings of cityscapes by Le Corbusier, or more unexpectedly in places such as Robert Venturi and Denise Scott Brown's examinations of Las Vegas; all of these undertakings contain images of dust. Dust seeps into the interiors of photographs in the architectural history books and, of course, into the interiors of everyday life. It's a form of subnature that gathers itself, heaping, sifting but also ready to be unleashed by far-off winds or a passing automobile. Dust is the result of natural decay in buildings, pollution from cars and factories, and the result of landscapes transformed by disasters. Unlike many other forms of subnature under review in this book, dust is always pervasive; though infinitesimal, it is never not there. Architectural historian Teresa Stoppani emphasizes this aspect of dust, noting that it is difficult to conceptualize dust as a truly transgressive element in architecture, something that crosses into architecture unexpectedly. She writes, "It does not transgress: it invades and pervades."[1] Stoppani sees dust as such a general part of everyday life that we must concentrate to even notice it.

Within architectural theory, dust often appears as a type of architectural background, an element that registers time by way of accretion. This is the type of dust we find in images of ruins by Piranesi, in particular his *Views of Rome*. In these images, ruins appear half buried, slowly taken over by dust. [**Fig. 5.1**] In this strand of thinking, dust might ultimately consume space but only over hundreds if not thousands of years. Accreted dust eventually registers as dirt, then soil, and eventually as earth itself. John Ruskin was probably the first architectural theorist to consider the temporal aspects of dust and dirt in depth; in his philosophy, dust was a type of historical record. To remove dust from buildings was, in some sense, to deny their own history. Ruskin lashed out at contemporary forms of restoration involving cleaning the exteriors of buildings, renewing them, but also eliminating their very standing as historically conditioned structures.[2] In these works and others, dust is largely innocuous; it registers its particular power against and over architecture, through neglect over expanses of time. Cultural theorist Georges Bataille provided a more recent conceptualization of dust. He claimed that dust was an uncanny register of time. Imagine, he asked, the thick blankets of dust covering the character Sleeping Beauty after one hundred years of sleep. She would no longer be an image of beauty but of

Fig. 5.1
View of Campo Vaccino,
Giovanni Battista
Piranesi, 1757

Fig. 5.2
Section of Mount Vesuvius emitting dust, fire, and smoke, drawing by Gabriel Pierre Martin Dumont, 1769

disgust. For Bataille, dust is what lies in waiting, ready to invade a world without human beings. Dust haunts space; by our understanding, it is a type of material future that awaits. Not only will we turn into dust, but without the presence of human beings to remove it, dust begins to dominate space.[3]

In contrast to the image of dust that registers time, there is another dust that most palpably emerges with sudden disasters or massive technological and ecological changes. This is a dust that is geographic and intense versus temporal and volumetric. This is the dust that the architect Gabriel Pierre Martin Dumont drew as erupting out of the crown of Mount Vesuvius—the famous volcano that buried Pompeii. In Dumont's image, a remarkable cross-section of the mountain, Vesuvius appears as a geological engine spewing smoke, dust, and fire. [**Fig. 5.2**] The destruction wrought by Vesuvius is but one of the myriad "dust disasters" that move through human history. In his examination of the social history of dust, historian Joseph Amato documented the different ways in which dust both causes and is a marker of disaster. In examining the Great Chicago Fire of 1871, Amato noted that it "poured such large quantities of ash into the sky that four days afterward the cinders reached the Azores in the middle of the North Atlantic."[4] The island inhabitants of the Azores considered the event a harbinger of an apocalyptical end. Amato goes on to cite another example of a geographic expanse of dust more familiar to North Americans; he writes, "dust imprinted the 'Dirty Thirties.' From 1932 to '38, the so-called Dust Bowl covered 150,000 square miles of the American plains, causing 60 percent of the people there to emigrate.... Dust found its way into people's beds and food; it tore at their skin and caused 'dust pneumonia.'"[5] Dust not only produced an environmental disaster, it became associated with the failures to coordinate modern development of land and settlement. [**Fig. 5.3**] Dust appeared as a vengeful force of mismanagement.

Within the above examples, dust appears as a force of nature, an element exaggerated by social forms of development but largely outside of society. Within a modern context, we begin to see an understanding of dust as a social form. This aspect of dust can be found in architectural thought in the writings of Le Corbusier. As we explored in the section on exhaust, Le Corbusier proposed towers that rose above the dust-ridden atmosphere created by urbanization. For him, the street was a space of dust, but unlike earlier architectural cartographers, he saw dust as a swirling menace rather than a benign register of time; dust was more alive. He wrote, "Dust, smells and noise stifle our towns of today." He proposed towers that worked against this dust-ridden street; they "are far removed from all this and

Fig. 5.3
Dust storm on the American Great Plains, 1939

Fig. 5.4
The dusty destruction of the Pruitt-Igoe Housing Complex, 1972

set in clean air amidst trees and grass."[6] In this new city of towers, Le Corbusier provided a refuge from the urban dust storm. The irony of his proposal was that within the context of modernity, buildings participate in generating urban dust storms as much as providing refuge from them. The destruction of buildings as part of modern and contemporary development emits enormous amounts of dust into the atmosphere. Within Paris, the dust generated by the destruction of buildings by Georges-Eugène Hausmann's planning projects was so intense that it spawned a new fashion. In the mid- and late eighteenth century, women walking down the streets and boulevards wore veils to alleviate the impact of dust spewing from demolition sites. These images of women in veils can be seen in etchings from street scenes and in several key impressionist works from the late nineteenth century.[7] The image of dust as a signifier of urban change continued in the paintings and drawings of New York City building construction in the early twentieth century. In these works, notably those of George Bellows and other contemporary realist painters, images depict workers laboring in swarms of dust. Demolition techniques, particularly those used to destroy concrete towers, produce such enormous amounts of dust that the blocks surrounding them must be evacuated before they are destroyed. [**Fig. 5.4**] In this latter aspect, as an atmospheric force, whether through disaster or not, dust not only represents a counter form of nature, it produces a new milieu, a new environment through the colonization of earth and air. Dust, from demolition and construction, blocks the sun, lowering the temperature of the spaces immediately beneath it by several degrees. It transforms the odor and, most disturbingly, the pH balance of air, and the dust produced from demolishing buildings contains alkali levels rivaling some the most powerful industrial cleaners. But in addition to all of these chemical and climatic aspects, dust is most often associated with dirt, soiling the surfaces and windows of buildings and coming inside through openings and passages.

As much as we try to control it, dust pervades us, and as much as it remains a nuisance—as a conveyor of destruction, hygienic obssessiveness, and historical change—dust in architecture contains a certain type of material and intellectual power that might be engaged, even as we try to wash it away. Such an engagement with dust might produce a negotiation that neither relinquishes space to dust nor rejects its prominence in urban life. By addressing dust, we can confront its particular physical and emotional power, its connotations as a historical register, and its existence as a form of matter that moves with calamities and in the most banal moments of time. Recent architectural projects spanning installations to preservation

Fig. 5.5
Observing the dust cloud hovering over the city, *New York, September 11, 2001, Four Days Later*, by Laura Kurgan, 2001

Fig. 5.6
Installation view of *Four Days Later*, by Laura Kurgan, 2001

work take on these aspects of dust. These understandings range from incisive social and political critique to more nuanced examinations of architectural maintenance; they explore how contemporary explorations of dust continue its position as historical marker, signifier of disaster, and as an object of compulsive hygiene. For example, in *New York, September 11, 2001, Four Days Later*, Laura Kurgan considered how the devastating attacks of September 11th were envisioned as a cloud of dust and debris. Using a photo taken by the Ikonos satellite, she argued that this type of orbital imager only reinforces our understanding of disaster as a register of the transformation of matter—a cloud of dust and smoke—versus a massive loss of life. [**Figs. 5.5 + 5.6**] The satellite image of the World Trade Center is the image of disaster within a society in which methods of surveillance provide our visual representations of cities. Kurgan laments that what is

Figs. 5.7 + 5.8 (opposite and right) Engaging with fears of dust in *Cloud*, by An Te Liu, Venice Biennale, 2008

missing in these images of dust, smoke, and debris "are the missing." This almost geological image of Manhattan enables us to distance ourselves from the full effects of the tragedy. In considering another recent examination of dust, we might recall that immediately after the attacks, households in Lower Manhattan were provided with air cleaners to both filter the atmosphere and to provide some sense of solace within an urban environment having unknown atmospheric content and potentially harmful effects. Such an environment is the context for the project *Cloud* by An Te Liu. This is certainly a different type of exploration of dust than Kurgan's, but it nonetheless relates to Kurgan's project through its focus on the monitoring and surveillance of space. In *Cloud,* Liu assembled numerous air cleaners into an apparatus hovering above an entryway to one of the gallery spaces in the Venice Arsenale. He wrote how this assemblage of cleaning apparati "wash, filter, ionize, ozonize, and sterilize our airspace, separating us from bacteria, allergens, germs, spores, dust and other bad things." [**Figs. 5.7 + 5.8**] He claimed that the project was not only a literal cleansing device but an urban representation: "a floating polis, perhaps of the future, but also recalling visions of futures past."[8] But the irony of this project is that the dust is simply moved, not removed. The idea is literally born out of the inner operations of each machine's dust-storage chamber. Our desire to clean results in the transference and relocation of dust. But returning to our original concept of dust as marker of history, we might consider one last project that represents a type of rapprochement with this substance—an experimental preservation project by Jorge Otero-Pailos, titled *The Ethics of Dust.*

Within an aluminum factory, he used a latex cleansing technique widely used in contemporary preservation for removing dust from the dirty surfaces of buildings. The pollution was lifted onto the latex and off the wall, where he hung "pollution casts" on a metal frame. [**Figs. 5.9–5.12**] The project is named after Ruskin's treatise on dust. Describing how this particular thinker inspired the project's development, Otero-Pailos wrote:

> Ruskin saw dust as an indication that materials were constantly changing: dust crystallized into stone, which then became dust again. He also associated different material states with different ages (youth, middle age, old age). Ruskin's thinking about dust made me think about pollution as a modern industrial material, and also about its connection to our conception of history.
>
> Arguably, pollution is like steel, glass, or concrete. But it is also unlike all other modern materials in that it is unintentionally produced. As it settles upon buildings, it assumes forms that are not intentional, but are still beautiful. The installation makes those unintentional forms

Figs. 5.9–5.12 (opposite, this page, and overleaf) Preserving dust as historical artifact, ***The Ethics of Dust***, by Jorge Otero-Pailos, Balzano, Italy, 2008

> visible by isolating them from the building. You can really see the patterns of the dust through the latex, especially when the sun comes through the windows behind. I was also interested to let people inhabit the space between the layer of pollution and the newly cleaned wall. I wanted to create a new experience of the relationship between the dust and the building, and possibly to open up a non-linear way to conceive of the changes that buildings undergo in time.[9]

Otero-Pailos's *The Ethics of Dust*, like François Roche's B_mu Tower, forces us to coexist with both the image and matter of pollution—a counterimage to most images of pollution in architectural discourse. This project enables us to see dust as a type of historical record, and it helps us to see that "cleaning dust" never implies the absence of dust. It reinforces a more tempered notion of dust as a register of time, keeping at bay the dusty environment's more troubling connotations.

Puddles

Although we often think of puddles as inconsequential, they appear in architectural history in prominent ways—in drawings of ruins, photographs of decaying buildings, and experimental designs that attempt to use water in provocative ways. Architects and historians typically draw or photograph these stagnant accumulations of water to represent specific forms of social transformation and loss. This use of the images of puddles connects eighteenth-century drawings of Roman ruins by architects such as Robert Adam with contemporary imagery of puddle-strewn modern buildings as in the photographs of Leonardo Benevolo. Curiously, these stagnant pools of water, once signifying society's vulnerabilities, appear to have disappeared in much contemporary work; like the eighteenth- and nineteenth-century hydraulic specialists who fundamentally viewed stagnant water with suspicion, contemporary architects continue this propelled vision of water in designs that emphasize movement and flow. But while images of aqueous flows are enticing, the seemingly pathetic image of the puddle, the trickle, or the leak has the capacity to represent social life in a more vulnerable, naked, and self-reflective state. In a few contemporary works we begin to see this slower, more stagnant image of water creep back in, suggesting a new type of socionatural interaction that architecture might initiate.

Some of the earliest images of puddles in architectural thought appear in the intensely archeological climate of eighteenth-century architectural theory. Architects Giovanni Battista Piranesi and Robert Adam both documented the ruins of ancient Roman reservoirs and drains in the town of Albano, a suburb of Rome. Their work was unusual in that despite representing Roman buildings in states of collapse, they conveyed the magnitude of ancient engineering. Piranesi in particular depicted many of the ruins at Albano with stagnant puddles and pools of water, as if to emphasize this civilization's eventual inability to manage water. In one image, we see an individual bent down in a former water storage building, inspecting a meager-looking puddle, and it reminds us of the space's former grandeur. [**Fig. 6.1**] Following his work at Albano, Robert Adam also conducted an important survey of the Emperor Diocletian's Palace at Split that included images of ancient waterworks. In one famous image, Adam drew the citizens of Split engaging with an ancient fountain that was still in existence in the city's periphery. The ground on which the fountain pours water is in a state of ruin, spilling water into an undefined pool filled with peasants and farm animals. [**Fig. 6.2**] This image of water depicts the ruined state of Split and

Fig. 6.1
A puddle in the drainage works at Albano, Giovanni Battista Piranesi, 1764

Fig. 6.2
"View of the Aqueduct which conveyed water from Salona to the Palace," drawing by Robert Adam, 1764

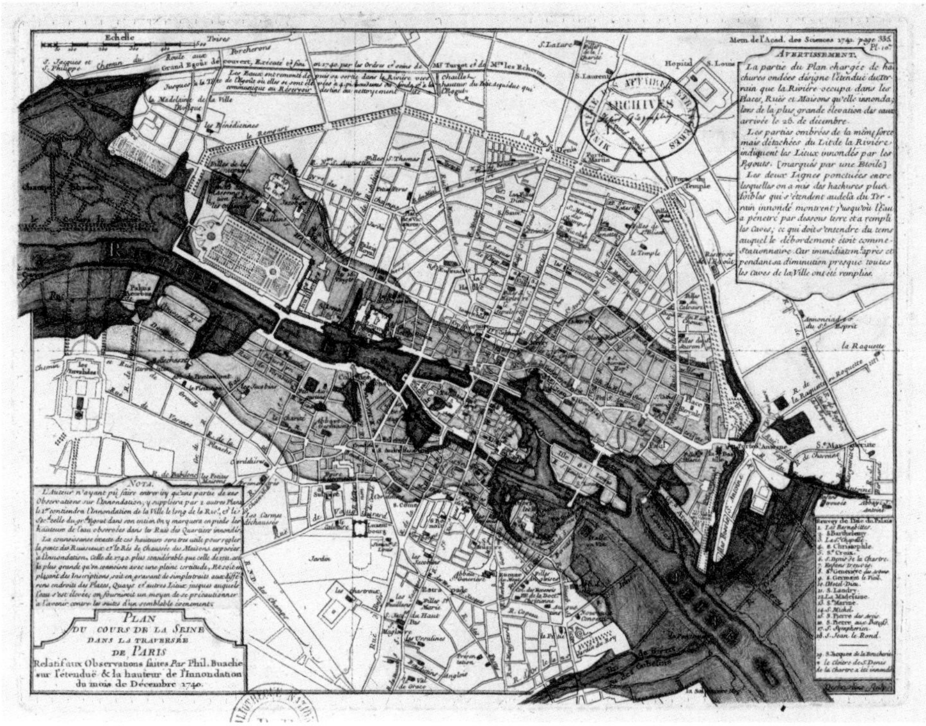

Fig. 6.3
Paris flood map, Philippe Buache, 1740

Fig. 6.4
Street section, Pierre Patte, 1769

the appropriation of an ancient aqueduct in contrast with the city's former negotiation of water and engineering.[1]

These architectural images certainly captivated a European audience on account of their drama, but they also, both directly and indirectly, informed the management of water in European cities. For urban managers and architects, the negotiation of water supplies and discharges became an important and growing concern within cities such as Rome, Paris, and London. Particularly significant were the problems faced by floods, sewage, and other undesirable waters that emerged from urbanization processes. The eighteenth century might be known as the century of floods in Western Europe, with significant events ravaging cities in 1727, 1733, and 1740. In Paris, the flood of 1740, chronicled in a map by the architect turned geographer Philippe Buache, destroyed property stretching far past the banks of the Seine. [**Fig. 6.3**] Buache's map, which charts the reach of the Seine and the large puddles and pools of water left as the waters receded, demonstrates the types of hybrid image-making that linked geological forms of cartography and the movements of bodies of water with more explicitly urbanistic forms of cartographic description, such as maps of city streets.[2]

The connection between ancient imagery and the above metropolitan concerns of flooding registers in several projects, particularly those concerned with river courses and roadways. Adam drew on his work at Split to develop new types of interactions between city and water in his Adelphi Terrace project along the Thames in London. The project provided a distinct boundary for the Thames and its upper avenues that, quite literally, recalled the sea wall of Diocletian's Palace. Piranesi's etchings of Roman ruins were distributed throughout Paris by architect and architectural theorist Pierre Patte. Influenced by Piranesi's etchings of Roman drainage works, the emerging French urban planning tradition, and the recent reconstruction work following the 1755 earthquake in Lisbon, Portugal, Patte developed an image for a proposed section of a street that articulated new ways of managing rain and wastewater. [**Fig. 6.4**] Patte's image is considered one of the first cross-sections to include both the neighboring buildings that define the street and the interconnected works of engineering that link interiors to the urban underground. In this latter role, Patte's drawing is also the first to articulate the fluid discharges of the city that arise within water and sewerage networks. Not only is wastewater conveyed from the building's interior to the sewer, but Patte considered the way that water running off a building's roof and onto the street might also be sent into the underground sewer. The drawing of the street section articulates a prominent space for the sewer itself, opening such conduits to a more strictly engineered interpretation.[3] This type of engineering and archaeological image advances the concept of the street as a strata of sloped and interconnected layers for effectively managing liquid waste streams. This notion of the street is maintained to this day in contemporary books on managing urban water.

If the engineered movement of water through and out of a city marked a civilization's height, then the breakdown of such a system also marked its vulnerabilities. While the eighteenth and nineteenth centuries provided the public with images of a fallen, grandiose ancient civilization, then in the mid- to late twentieth century, a new type of image of the wet ruin emerged to articulate the potential downfall of modern cities. Beginning in the 1960s in newspapers, film, architectural history, and design, we begin to see images of the undrainable city as a type of counterimage to Patte's modern street concept. What differed from earlier images was not only that these city scenes appeared unmanaged and unengineered, but that the waters inundating the modern city rained from above and surged from below. From the mid-1940s to the mid-1980s, images of burst water mains in New York City and other cities such as Chicago and Boston signified the new, unnatural discharges in which urban subnature bubbled up and into the city's spaces. In one image of New York City's 42nd Street in 1977, we see the city as it might appear after a rainstorm, but the water is coming up through mains below the street. [**Fig. 6.5**] The signs and streetlights of this already notorious section of the city are reflected in this image of a street—a type of antiboulevard against the drawings of Patte. Within works of architectural history and design, images of puddles and undrained spaces highlighted the fragile nature of modern architecture as it interacted with the environment over time. Crumbling puddle-ridden photographs of famous works of modern architecture appeared prominently in Leonardo Benevolo's *History of Modern Architecture* (1977). [**Fig. 6.6**] If the puddle conveyed urban decay in the images of New York City, in Benevolo's photographs they communicated a type of realism. For him, the puddle demonstrated how modernist architecture is "lived" versus represented in a frozen state in most histories.[4] Ultimately, the puddle acts as a provocation against the representation of modern architecture in photographs of weatherless contexts. This concept of the puddle also appears in the commissioned photographs of the Economist Building Group, by Alison and Peter Smithson. [**Fig. 6.7**] The photographer Michael Carapetian chose to shoot the buildings after a rainstorm, and the puddle-strewn plaza invoked a new image of modernity in architecture that appears far less

Fig. 6.5
Nighttime flood in Midtown Manhattan, 1977

Fig. 6.6
Puddles in front of the Bauhaus, Leonardo Benevelo, 1970s

Fig. 6.7
Man on the plaza of the Economist Building Group, photograph by Michael Carapetian, 1964

staged, or what the Smithsons might term "authentic." It is quite close in sentiment to the photographs by Benevolo, in that the buildings appear somewhat weak in their engagements with nature. Carapetian's "wet" photos became the dominant representation of the Economist Building Group, reproduced in hundreds of articles and books discussing the complex. Like Benevolo, Carapetian sees the puddle as a provocation of both the experience of a building and the representation of it as a historically conditioned object—the puddle is an avenue to a slightly disturbed realism. This notion of the puddle as the "real" continues to the present, perhaps most literally in the WOS8 project, in which NL Architects wrapped a utilitarian building (a heat exchanger) in a completely detail-less rubberized shell containing spaces for people to climb and play ball, for birds and bats to nest, and for rainwater to linger. The architects intentionally designed the building's skin to enhance the wet climate of the Netherlands—an aspect of the building's perverse sense of contextualism. [**Figs. 6.8–6.10**]

While many contemporary architects are intensely focused on enacting images of nature in flow, there has also been a renewed interest in images of stagnancy (described above) and nuanced expressions of drainage that question its association within a well-managed society. More novel architectural representations of the social impacts of water can be found in a series of projects designed within the last five years by Jason Johnson/Nataly Gattegno, Philippe Rahm Architects, and Thom Faulders Studio, among a small handful of others. These projects extend the "wet" architecture of the Smithsons, NL Architects, and others interested in water as something that provides a novel experience and that can be positioned as a type of critical evaluation. For example, in the Super Galaxy: NYC Tropospheric Refuge project by Jason Johnson and Nataly Gattegno, the architects imagine the upper floors of a highrise building inundated with water and urban wildlife. A robotically controlled floor adjusts its contours to produce puddles, wet zones, and large pools that might be appropriated by groups of people or migrating fowl. [**Fig. 6.11**] Above the wet surface, another area rises and lowers, producing a series of rooms. The architects imagine the entire surface as adjusting in response to natural, social, and cultural data. They write:

> Super Galaxy is an architectural system saturated in atmospheric and electronic phenomena.... As it fluctuates between states of varying coherence (solidity, liquidity, and gaseousness) its inner structure maintains an invisible, yet definable pattern. It is a responsive system capable of dynamically interacting with its surroundings on many levels. It is in a constant state of motion as it calibrates and recalibrates relative to

Fig. 6.8
Drainage opening of the WOS8 project, by NL Architects, Utrecht, the Netherlands, 1999

Fig. 6.9
The wet skin of the WOS8 project, by NL Architects, Utrecht, the Netherlands, 1999

Fig. 6.10
WOS8 project, by NL Architects, Utrecht, the Netherlands, 1999

> both real-time global datasets (weather, pollution, warfare, etc.) and local datasets (desired micro-climates, heat exchange, light, and sound).[5]

If the weather changes, the pools disappear. During a warm month or in the event of a large migration of birds, the building's floor can adjust into an entirely flooded interior landscape. Where NL Architects reinvigorate Benevolo and Carapetian's notion of the puddle as "the real"—an imagined counterpoint to the apparent resistance of modern architecture to this particular socionatural milieu—the puddles of the Tropospheric Refuge suggest that the puddle is no longer a counterpoint to architecture but intrinsic to it. [**Fig. 6.12**] Neither traversed nor accidentally encountered, the puddle is utilized as an existing form of urban nature. A slightly more subversive notion of the flooded and leaky interior appears in the Mollier Houses by Philippe Rahm. In developing the concept for this house, Rahm notes that the architectural interior is constantly inundated with invidious forms of moisture. He writes:

> An occupant of an indoor space produces water vapor, not in a constant manner, but according to the primary activity to which each room is dedicated. The presence of water vapor in the air originates naturally from respiration and hot water usage, leading to risks of condensation and damage to the construction. While today, the only solution to excess interior water vapor is the common use of technical ventilation systems,

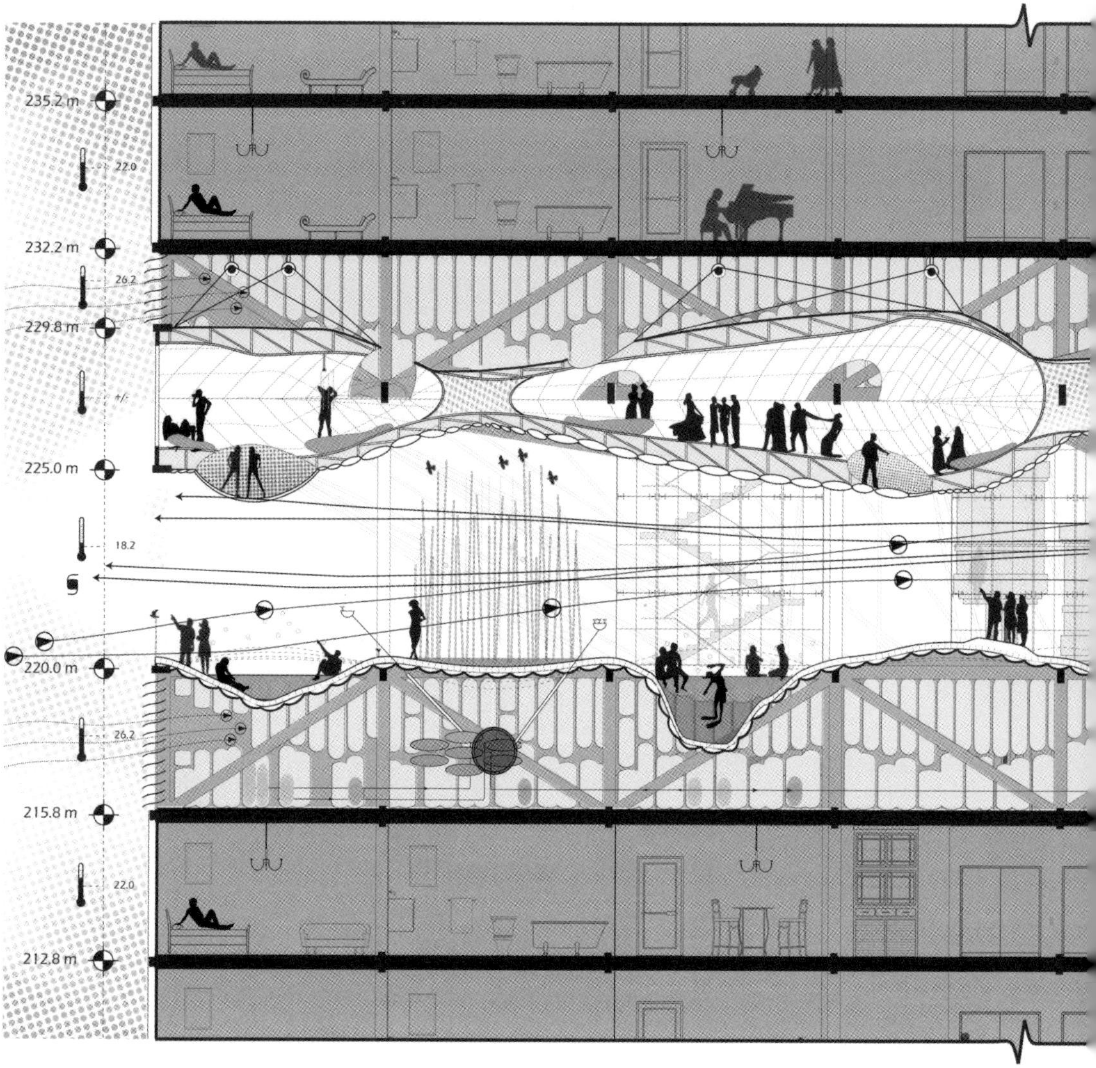
235.2 m
22.0
232.2 m
26.2
229.8 m
+/-
225.0 m
18.2
220.0 m
26.2
215.8 m
22.0
212.8 m

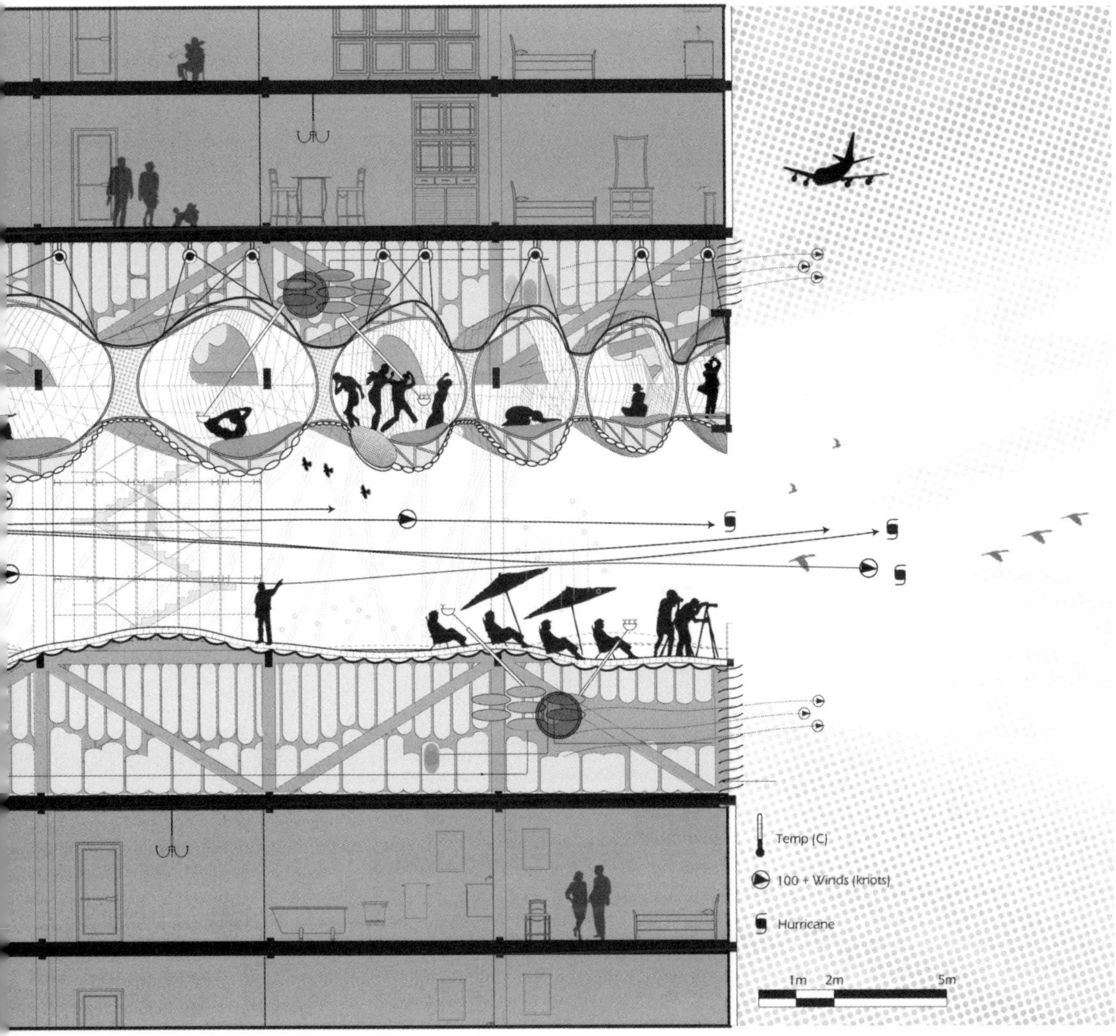

Fig. 6.11
Puddles as part of a large public zone situated within the central floors of a skyscraper, section diagram of Super Galaxy: NYC Tropospheric Refuge, by Jason Johnson/ Nataly Gattegno, New York City, 2005

> we propose in this project to shape the space in relation to the water vapor, in order to inaugurate a profound and complex relation between the inhabitants, their bodies, and the space according to its physical and chemical characteristics.[6]

Rather than simply removing or remediating this interior-generated moisture, Rahm proposes structuring a building to harness it—developing a sequence of spaces that move from dry to moist to a space of 100 percent humidity. The house's plan is based on the Mollier humidity diagram, with Rahm suggesting that the most humid space contains a flooded room, with its humidity spreading into adjacent areas of the house. [**Figs. 6.13–6.16**] Finally, where the above projects bring pools and puddles of water into architecture, an altogether different concept of inundation emerges in the project Deformscape by Thom Faulders. Deformscape's intensely warped surface returns us to the image of architecture as drainage, but it advances the need for drainage to the point that it overwhelms our ability to imagine inhabiting its surfaces. Once again, drainage appears as a representation, but where the above projects advance a renewed appreciation of the stagnant pool in architecture, Faulders completely rejects the traditional relationship between water management and urban inhabitation. In his proposition, water management is in full effect, but it acts against the very idea of usability. [**Figs. 6.17 + 6.18**]

The aforementioned contemporary projects, taken together with historical examples, suggest that the concept of stagnancy in architecture might be reconsidered in an era in which flow appears as the operative concept for urban interchange. As stagnancies literally produce images of social constructs in their glassy wet surfaces, the concept of the stagnant suggests that all efforts of nature management in cities contain representations, and that these representations demand renewed and intense theorization.

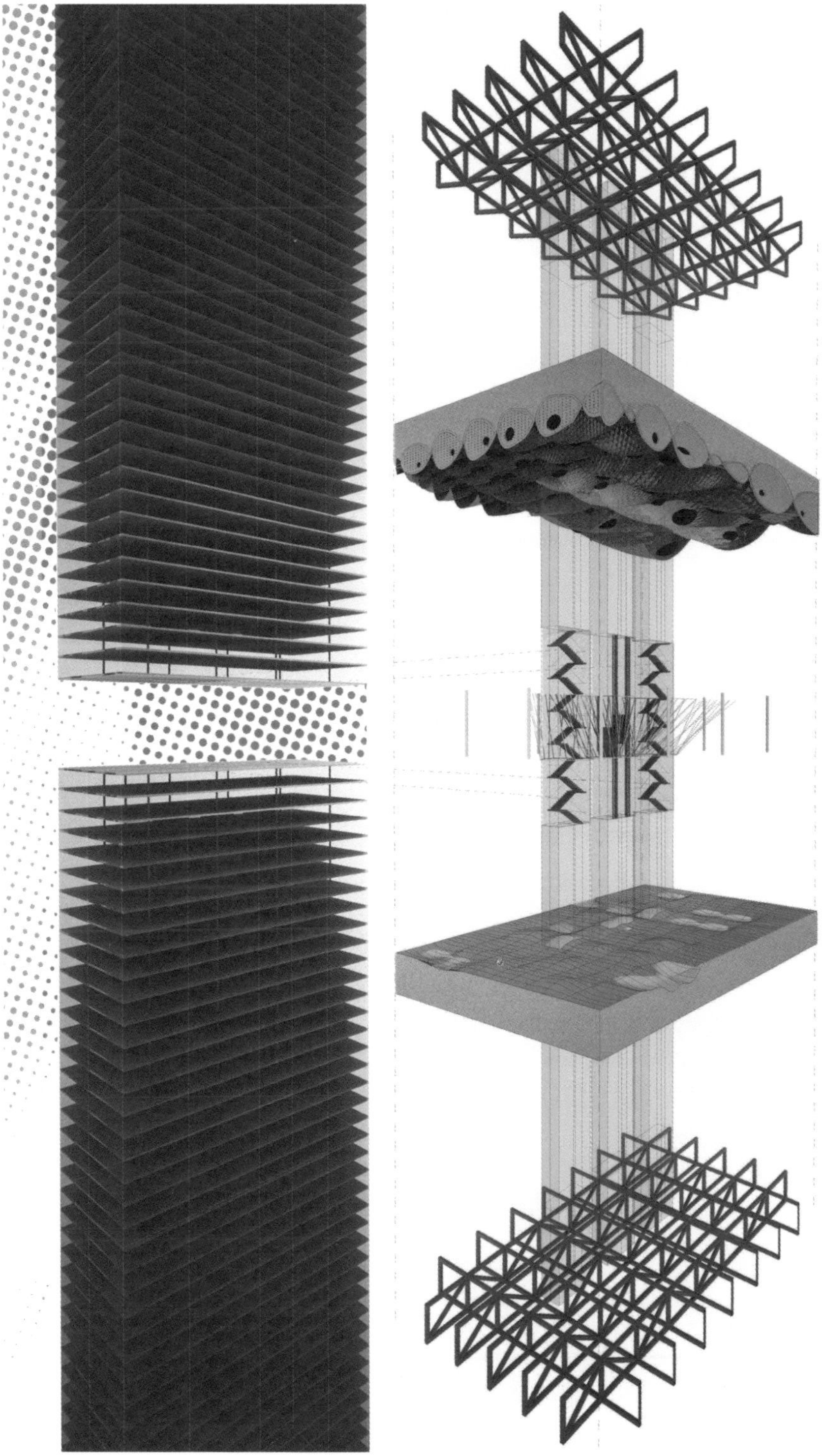

Fig. 6.12
Exploded axonometric diagram of Super Galaxy: NYC Tropospheric Refuge, by Jason Johnson/ Nataly Gattegno, New York City, 2005

Figs. 6.13 + 6.14
Exterior and interior perspectives of the Mollier Houses, by Philippe Rahm Architects, Vassivière, France, 2005

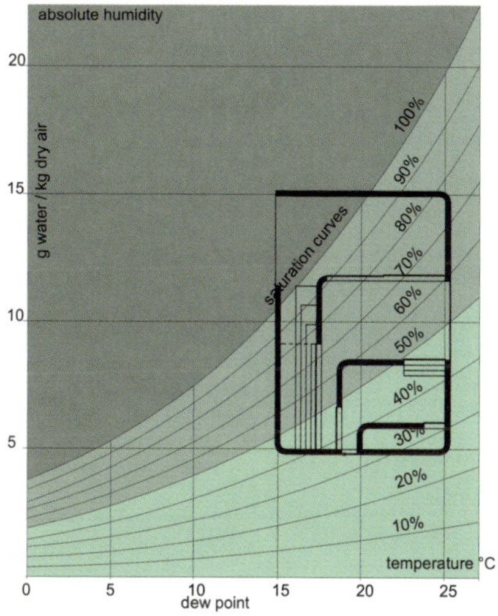

Fig. 6.15
Plan and humidity diagram, Mollier Houses, by Philippe Rahm Architects, Vassivière, France, 2005

Figs. 6.16
Interior of bathroom, Mollier Houses, by Philippe Rahm Architects, Vassivière, France, 2005

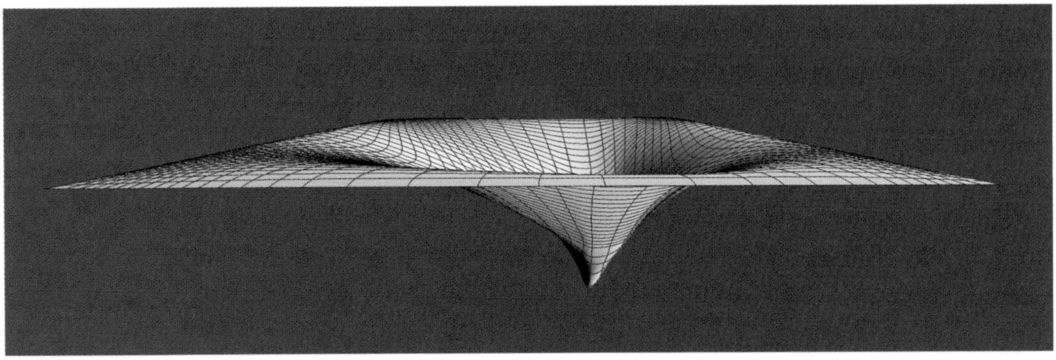

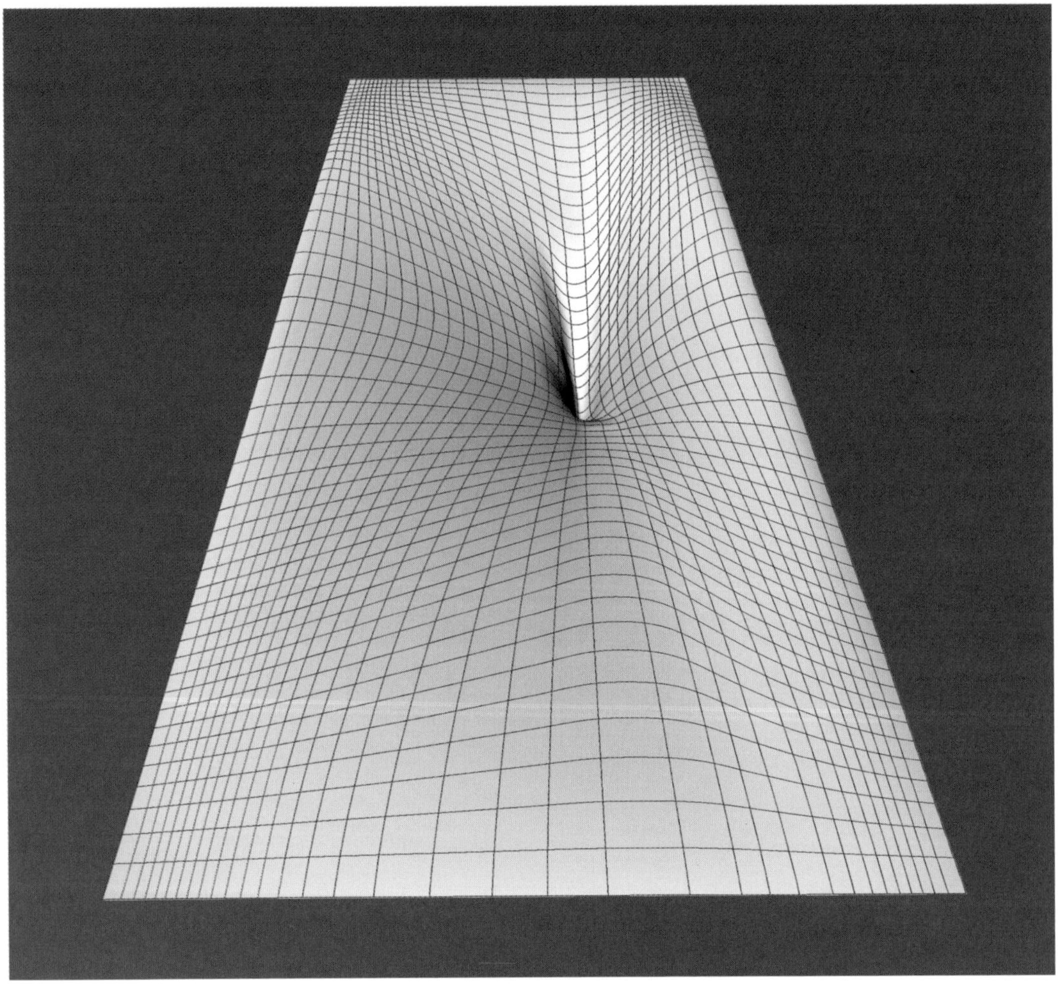

Figs. 6.17 + 6.18 (above + opposite) Deformscape, by Thom Faulders, San Francisco, California, 2009

Mud

Mud, the viscous mixture of soil and water, appears in architecture in several ways. It is one of the original materials of architecture, central to concepts of the primitive in architecture. Mud is also a type of unstable ground that must be overcome in the construction of foundations. In the development of modern cities, rather than discussing mud as something wanted or desirable, it was identified as the product of poor drainage and ineffective engineering. Like puddles, mud often signifies a type of failed engineering, but it also takes on a larger metaphorical dimension in architecture. As an aspect of primitive architecture, mud was once associated with a lack of cultivation—a type of degenerate architecture. As a viscous substance, it operates against modern concepts of circulation; it slows the city down, like slush. As something dark and often very deep, mud contains notions of history; it embalms things and contains the past. Like other subnatures in this book, the development of ideas about mud is intimately related to the rise of modern economies, industrialization, urbanization, and the birth of the modern state. Mud has run counter to virtually all of these formations. When economic prospects went bad, in a sense, they turned into mud (or dust); urban routes were slowed, and mud infiltrated all manner of economic enterprises, from farms to mines. But within the formation of large nation states, mud's role was most peculiar. It often became both the matter and metaphor that states attached to their colonized subjects' appearances, building practices, and cities.

While the rise of all Western nations involved social negotiations of mud, we might focus our consideration within the modern, industrial, and colonial rise of the United Kingdom. Within this context, the conflict with mud is rife with historical imagery and concepts that extend from the eighteenth century to the present. Mud became associated with urban decrepitude, the primitive, and national liberation. Within England and London, more specifically, we see some of the first architectural reflections on the history of mud construction and the first massive engineering efforts to remove mud from the city, as well as an embrace of the substance by former colonized subjects of the British Empire. Within this context, mud is a thoroughly countermodern material, but it makes a surprising reappearance within the practices of those who wished to unshackle themselves from the very forms of governance that implicate mud with degenerate qualities and those who wish to critique the material prejudices of modern Western societies.

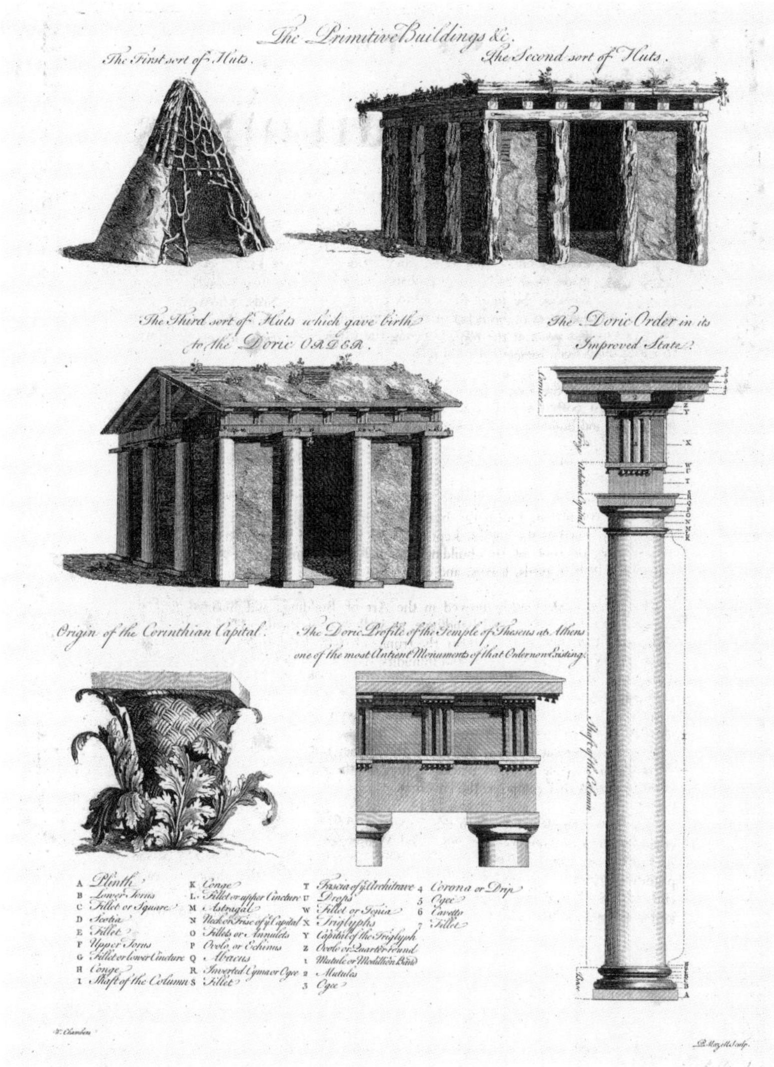

Fig. 7.1
Primitive mud huts predating architecture, from *Treatise on Civil Architecture*, William Chambers, 1759

For example, in the eighteenth century, during the development of a distinctly British architectural theory, William Chambers, architect to the King of England and surveyor of the monarch's engineering works, provided one of the more cogent narratives of the role of mud in architecture.[1] In a drawing for his architectural treatise on the development of architecture, Chambers illustrates a series of building types: a house of wooden beams and mud, a house of wooden beams and plaster, and finally a stone building. [**Fig. 7.1**] In this drawing, we see architecture as a process advancing toward refinement, and this refinement is partially defined by a distance from primitive forms of mud-and-stick construction. Chambers was, in fact, the first architect to use the word *primitive* in his writing.[2] In his drawing, the mud hut is firmly placed in a lower aesthetic position compared to the others. This concept, fully illustrated and emphasized by Chambers, extends and elaborates on theories of Vitruvius, who wrote that early structures were largely dependent on the materials at hand, versus those that could be cultivated through larger geographical developments of territories. Vitruvius described two primitive houses to articulate his idea—the House of the Colchians and the House of the Phrygians. The Colchians had abundant timber and built their houses out of wood, while the Phrygians had limited timber and instead created large mud embankments that were roofed with sticks and mud.[3] Within Chambers's work and in Vitruvian theory more generally, mud buildings were among the origins of classical architecture, but we should also understand that within an expanding colonial milieu, mud architecture has been the material of "savages" and the poor. The house illustrated by Chambers is close to that of the Phrygians described by Vitruvius, but it also invoked the types of structures colonialists were encountering in their expeditions abroad.[4] We might add that the slightly less primitive structures illustrated by Chambers were very close to the types of buildings constructed by English, Irish, and Scottish peasants. In the late nineteenth century, illustrations of indigenous mud dwellings, both within architecture and outside of it, illuminated the experiences of English colonial expeditions while also cementing the notions that colonized subjects produced structures that were part of architecture's past. We might see this as the groundwork for more explicit socially Darwinist ideas of the early twentieth century, at which time the genealogy of history became an act of social denigration and in which notions of "primitive" and "barbaric" were more strictly associated.

Chambers was both an architectural theorist and more significantly the architect to King George III, who effectively expanded British interests in Ireland, the Caribbean, and India. The wealth brought back from these imperial conquests enabled the reconstruction of the empire's capital city, London. For George III, Chambers built the Somerset House, one of the largest urban plans realized in the city in the late eighteenth century. The Somerset House, a series of governmental offices, originally bordered the muddy banks of the Thames, one of the chief sites where various improvements were proposed and where conceptualizations of urban mud were formed. [**Fig. 7.2**] Both the Somerset House and the Adelphi Terrace (discussed in the section on puddles) provided direct access to the Thames and hardened the river's muddy edge. The waters of the Thames ebb and flow with tidal cycles, and as the tide moved out, the muddy banks were often exposed in the city. Such middle zones, between the hard

Fig. 7.2
The Somerset House rises from the muddy banks of the Thames River, designed by William Chambers, 1790s

pavement and the flow of the river, were considered sites of urban alterity. In several instances, mud was associated with specific forms of disease. Many believed that the corpses of animals that washed up or the "river slime" that mixed with the mud were the causes of cholera. The mud also contained the city's past—bits of old ships, crockery, and sometimes valuables appeared as the tide receded. The harvesting of objects from the Thames's banks was the work of so-called mudlarks. Often, these young boys were impoverished scavengers that searched the banks in teams for potentially usable, if not valuable, objects. Their efforts were mirrored in other forms of foraging, such as that of the pure-finders, who scoured the sewers for pure manure (for agriculture) and the toshers who searched sewer drains for coins and lost valuables.[5] [Fig. 7.3] The muddy banks of the Thames presented the city with a marginal space in the very center of the city, and it is this space that would become the focus of engineering efforts to restructure the urban environment.

Although London never experienced the wholesale reconstruction of urban space, as in the Paris of Haussmann, by the 1860s several select architectural and engineering projects were designed to both rationalize urban space and eliminate marginal zones such as the Thames's mudflats. In concert with the Somerset House, Adelphi Terrace, the redevelopment of Bethnal Green, and the

Fig. 7.3
Mudlarks on the Thames, circa 1900

Holborn Viaduct, engineers in London proposed the development of an embankment that would reconstruct the city's muddy border into an image of rationalist engineering. Designed by Joseph Bazalgette, the Thames Embankment involved the reconstruction of several sections of the Thames—most prominently the area surrounding the Somerset House. In his design for the three-story structure, Bazalgette included spaces for sewer and gas lines, locomotive travel, and paved roadways for carriages and pedestrians. [Fig. 7.4] The entire structure was finished in 1874, its construction occupying most of the previous decade. Where there was once a space of viscous urban nature, the Embankment offered an urban flow in which forms of travel and waste management were equated.[6] In developing a new thoroughfare in London, the Embankment all but eliminated the old muddy shore, replacing it with an imposing edifice and sheer stone walls. Ultimately, the Thames Embankment not only represented the exertion of control over urban space but also over a far larger socio-natural geography. A painting by John O'Connor, completed in 1874, demonstrates the underlying concepts. Titled *The Embankment*, it shows an image of Queen Victoria as the Empress of India standing on the terrace of William Chambers's Somerset House looking over a regiment of troops marching through this new urban space. As Lynda Nead notes, the painting's sweeping view and militarist sentiment is a "vision of London not as metropolis but as capital of an empire."[7] Within the larger context of England's expanding reach, the Thames Embankment not only embodied control over the nature of the city, but control over the natures encountered throughout the British Empire.

Projects such as the Thames Embankment in London were repeated throughout Western and colonial cities in the late

Fig. 7.4
Cross-section of the Thames Embankment, 1867

nineteenth and twentieth centuries. Massive paving projects were undertaken to remove muddy pockets from the centers of cities, and drainage efforts eliminated mud from most public parklands. The appearance of mud became emblematic of urban engineering failures and impending pathologies.[8] In colonized states, European building practices that employed concrete, steel, and glass replaced many traditional forms of construction. [Fig. 7.5] Mud, as matter and metaphor, largely disappeared from the future imagination of city and nation. By the mid-twentieth century, it had become a marginalized aspect of architectural practice, appearing in a few tracts on *pisé* or adobe construction in American and European regionalist practice. While the absence of mud in European cities reflected the socionatural values of metropolitan capitalism, in the post–World War II era, a resurgence of interest in mud by former colonized subjects emerged as a type of counterpractice against European interests. For example, in the postcolonial context of Egypt, the architect Hassan Fathy rejected the concrete, steel, and glass architectural language brought to Egypt by Europeans. Fathy theorized that the best construction material for the emerging Egyptian state would relate to traditional forms of construction and materials adapted to contemporary forms of planning. For these reasons, Fathy embraced mud construction, arguing that it was literally free and that it was best suited to the hot climate. Reflecting on the use of mud, Fathy wrote:

> Surely it was an odd situation that every peasant in Egypt with so much as an acre of land to his name had a house, while landowners with a hundred acres or more could not afford one. But the peasant built his house out of mud, or mud bricks, which he dug out of the ground and

Fig. 7.5
Nigerian man building a mud house, 1950s

Fig. 7.6
Mud architecture in Egypt, Hassan Fathy

> dried in the sun. And here, in every hovel and tumbledown hut in Egypt, was the answer to my problem. Here, for years, for centuries, the peasant had been wisely and quietly exploiting the obvious building material, while we, with our modern school-learned ideas, never dreamed of using such a ludicrous substance as mud for so serious a creation as a house. But why not?[9]

Beginning with designs in 1937 and after several experimental developments were completed years later, Fathy developed mud construction techniques that employed the material for the sake of impoverished rural populations. His mud architecture was characterized by its use of vaulted construction techniques, domes, and often spacious courtyards. [**Fig. 7.6**] He conjoined materials and subjects in ways that continue to inform contemporary architectural engagements with marginalized socioeconomic groups, as in the case of the work of Samuel Mockbee in the southern United States, and his experiments have been widely lauded and critiqued.[10] But more significantly, Fathy's work provided a material-based construction language for a postcolonial approach to architecture.

As novel as Fathy's efforts were in developing techniques for unshackling nations from colonial economies, he never fully untangled the range of implications associated with using mud or any other manner of "poor" material. Such notions were not possible within a modern episteme that emphasized material performance and economy over self-reflective representational concerns within particular construction materials. Fathy's late-modern approach to the use of mud has continued to the present, with architecture that utilizes this material for economy, naturalism, and thermal efficiency. But the impression that mud is a substance that contains provocative historical characteristics has mostly disappeared. In many contemporary projects, mud is quite literally dry.

We begin to see something that explores a more expanded idea of mud in a recent project for the Venice Biennale. In 2000, the nation of Ireland was invited to develop their first pavilion for the biannual event. For the pavilion, Tom dePaor Architects utilized mud from Irish bogs.[11] In doing so, their embrace of mud fit squarely within contemporary conflicts of Irish national consciousness. The Irish bog covers roughly a sixth of the island, and it is used as a source of agricultural nourishment and as fuel burned in power plants and domestic furnaces. But the bog is also a filthy place, often associated with Irishness in denigrating ways. The Irish poet Seamus Heaney wrote a series of poems in which the bog is described as "Earth-pantry, bone-vault, / sun-bank, enbalmer / of votive goods / and sabred fugitives."[12]

Fig. 7.7
Irish Pavilion, by Tom dePaor Architects, Venice Biennale, 2000

Fig. 7.8
View from the interior of the Irish Pavilion, by Tom dePaor Architects, Venice Biennale, 2000

Fig. 7.9
Irish Pavilion, by Tom dePaor Architects, Venice Biennale, 2000

Fig. 7.10
Detail of bog bricks, the Irish Pavilion, by Tom dePaor Architects, Venice Biennale, 2000

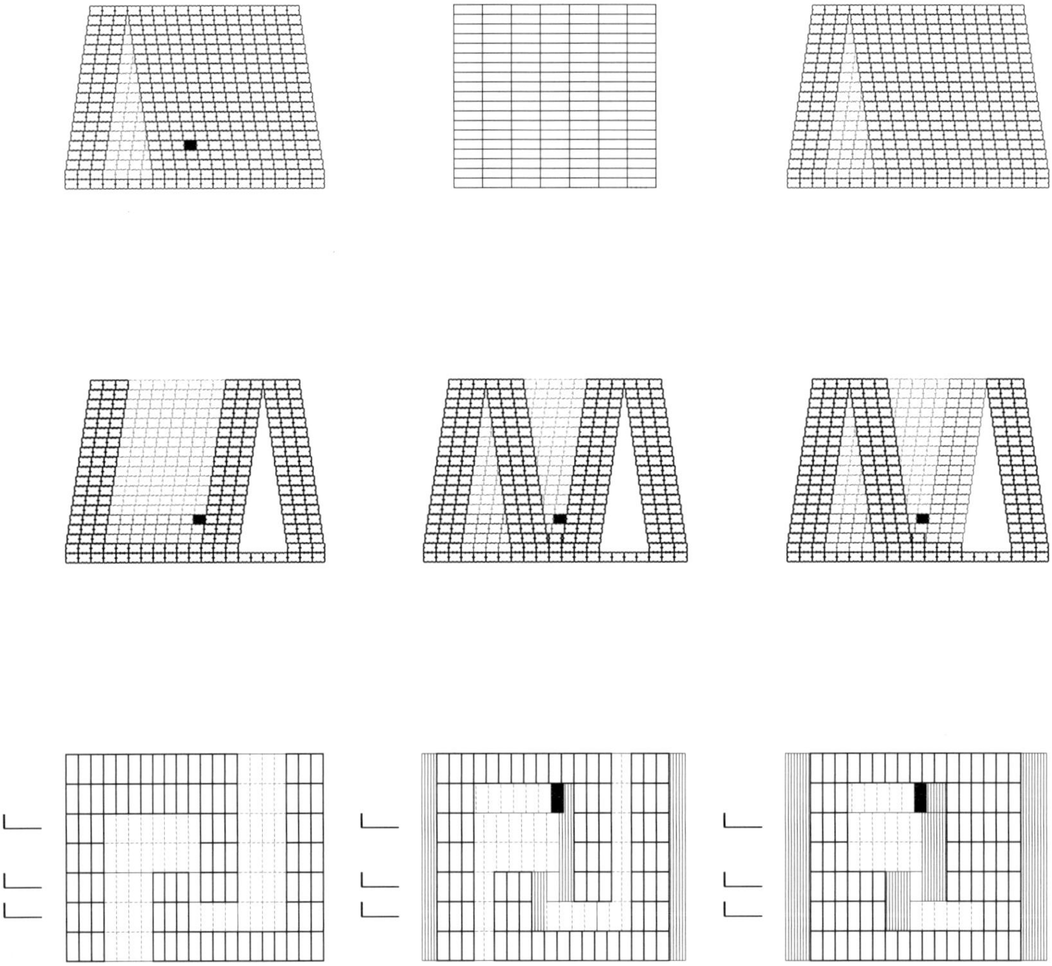

Fig. 7.11
Plans, sections, and diagrams of the Irish Pavilion, by Tom dePaor Architects, Venice Biennale, 2000

In Heaney's imagination, the bog functions in ways similar to the American frontier—as a boundary and a space of reinvention. But where the American frontier promises endless expanse, the bog offers mysterious and muddy depths. In their design for the pavilion, Tom dePaor Architects embraced the bog's varied associations. The project simultaneously invokes the type of mud construction denigrated by British commentators, while it negotiates mud's history as a marginalized, viscerally inappropriate material for a city. In the pavilion, the mud building is presented as a wet collective entity containing matter far exceeding its immediate appearance. The pavilion uses over twenty-one tons of peat, which further contains approximately 2,500 tons of water and sixty-three kilograms of sulphur. The dark-black bog bricks, with their moist sulphurous content, emit vaporous odors of the earth while also serving to cool the interior of the building. Mud confronts the space it houses through its dark, viscous, and wet character, but equally significant is the way mud confronts its historical condition when appropriated as an expression of statehood. [**Figs. 7.7–7.11**] Yet in this reappropriation, the architects further complicate mud's historical role. In this context bog mud suggests an image of something provisional, of an era in which the state as a cohesive concept of peoplehood appears to have lost its natural foundation. When the Biennale was over, Tom dePaor Architects oversaw the shredding of the pavilion, and they released this bit of symbolic Irish earth into the city of Venice.

Debris

The term *débris* originated in eighteenth-century France to describe a type of broken, scattered substance that was once part of a building or standing structure. The term differed from the French words *moellon* or *décombres*, both translating as "rubble." *Moellon* refers to rough stones pulled from quarries or used for paving roads, and *décombres* describes the wreckage from building demolitions. Within early-modern and modern French architectural writing, authors often used the word *débris* to describe the scattered and atomized remains of structures that had been leveled by cataclysmic events such as war or natural disasters. In contrast, other terms describing rubble suggested something larger, potentially salvageable, and local in terms of its distance from the building that it was once part of.[1]

The emergence of the term *débris* coincides with two important architectural developments. The eighteenth century witnessed the increased use of gunpowder in European warfare and the documentation of its effects on architectural structures, as well as, more generally, an increase in the archaeological documentation of tiny fragments of destroyed structures. Unlike the study of ruins during the Renaissance, the latter method differed in that it involved taking in the totality of bits that once composed earlier structures. An examination of *débris*, in this sense, was different from studying the architectural fragments of ruined sites, which generally focused on former building elements, apart from surrounding remains. A study of an architectural fragment could refer back to some physical whole—a column, an architrave, perhaps an entire structure, but debris was often referred to as a collection of unrecognizable matter. Debris takes in the total spatial transformation wrought by violence and disaster, and it speaks of the ways that destroyed structures transform their surroundings.

Discussions of architectural debris—as both the remnant of destructive events and as a territory of fragments—can be traced to the eighteenth century, specifically as an aspect of the French archaeological project in architectural theory. Architectural theorist Julien-David Le Roy was one of the first to provide debris with a specifically architectural visual character. Le Roy was among the first Europeans granted entry into the Ottoman-controlled regions of Greece. Prior to Le Roy's expedition, knowledge of Greek structures was limited to the writings of Vitruvius and surveys of the surviving temples at Paestum in Italy. Le Roy's drawings of Greek structures provided the groundwork for two movements in architecture: the Neoclassical movement (the adoration of a simplified Greek

Fig. 8.1
The scattered fragments of the Parthenon, after a drawing by Julien-David Le Roy, 1758

classicism) and what would eventually become the Picturesque movement (emphasizing the gradual decay of structures and their integration into natural surroundings). The slow creep of nature on architectural structures underscored time and the impact of history on our experiences of buildings. These contributions have been well documented by contemporary historians, but other significant ideas moving through Le Roy's images of Greek ruins have remained obscure. Although many of the structures he examined had reached their current state as the result of the slow creep of time and nature, his image of the Parthenon represented a singular human-caused cataclysmic event. It is somewhat deceiving, as he was not examining a well-aged, slowly decaying building but the results of an approximately eighty-year-old attack by Venetian forces that had exploded the ancient structure.[2] Le Roy's image of the Parthenon, with its side blown open and the resulting architectural fragments scattered in front of it, provides an evocative image of debris as the product of an act of violence. [**Fig. 8.1**] In this image, we are witnessing the destruction of an important ancient artifact and the resulting transformation of that artifact's surroundings. This is far more than a Picturesque dialectic between an ideal building type and the onrushes of time. Coursing through Le Roy's image is the notion that in one flash of a moment the distinction between social creation and nature are atomized.

The idea of destruction implicit in Le Roy's studies of debris becomes much more explicit 100 years later in the work of another French architectural theorist and inspector of ruins, Eugène Emmanuel Viollet-le-Duc. In his examinations of important French monuments, the architect, theorist, and preservationist explored the remains of numerous buildings throughout the state. Where Le Roy provides us with an emotive and atmospheric concept of debris, Viollet-le-Duc portrays debris as part of a rational engineering project of warfare and systematic destruction. In his book *Annals of a*

Fig. 8.2
Images of debris from *Annals of a Fortress*, Eugène Emmanuel Viollet-le-Duc, 1872

Fortress (1872), Viollet-le-Duc explores the numerous sieges that befell a French stronghold. He uses the term *débris* consistently throughout the text to capture the character of the physical matter produced by the attacks and the sense of violence against the inhabitants of this place.[3] The debris that Viollet-le-Duc inspected in his examinations of French historical sites is imagined as the register of battering rams and the work of modern artillery. As in Le Roy's images, Viollet–le-Duc's book portrays the ground as a site filled with shards, stones, and other material remnants. [**Fig. 8.2**] From both Le Roy and Viollet-le-Duc we realize that debris is matter that remakes the ground in images of violence. Debris does not talk about decay as a result of forces of nature but due to an incident, a cataclysmic socionatural event.

Where Le Roy and Viollet-le-Duc developed some of architecture's earliest images of debris, something we might term an architectural theory of debris did not emerge until eighty years later, in the mid-twentieth century's aftermath of massive warfare at a global scale. Immediately after World War II, architectural theories of debris emerged throughout Europe as well as within Japanese architectural culture. This was primarily in response to the destruction of European and Asian cities by the Allied and Axis powers. While early-modern and modern wars in Europe and Asia unleashed incredible destructive forces on cities; during World War II, for the first time

Fig. 8.3
Debris-ridden Neighborhoods after the bombing of London, 1940

in history, enormous cities in Europe and Asia were flattened, transformed into little more than debris sites. [Fig. 8.3] Architects reacted to the debris-laden cities of World War II in a number of ways. For the European CIAM group (the International Congress of Modern Architects), the ruined state represented opportunities for reimagining cities as blank slates, obliterated of their premodern histories. Some architectural thinkers, such as Ludwig Hilberseimer, wished to abandon cities altogether. Hilberseimer became concerned with the increasingly catastrophic nature of modern warfare, particularly the effects of debris from nuclear fallout.[4] But for another group of postwar architectural thinkers, the ruined sites of European and Asian cities were opportunities for reflection on the thing itself—the massive accumulations of debris that these cities had become.

In England, the New Brutalists, centered on the work of Alison and Peter Smithson and their larger Independent Group of architects, artists, and designers, sought an "authentic" architecture focused on the everyday experiences of postwar urban life. In post–World War II European cities, such an authenticity involved acknowledging the detritus that littered bombed and shelled urban centers. In the Smithsons's work, debris became a type of authentic nature that contrasted with to the green parkways and fieldscapes of other postwar architects and planners. In their *Patio and Pavilion*, the constituent elements of the heavily bombed sector of London's East End become

Fig. 8.4
Excavated after an atomic holocaust, the Patio and Pavilion installation, by Alison and Robert Smithson, 1956

Fig. 8.5
Piles of debris in the Robin Hood Gardens, by Alison and Robert Smithson, 1970

the materials for a new type of spatial construct. Formerly functional objects—bicycle wheels, tools, and other forms of rubbish—are scattered around, beneath, and above a simple shacklike pavilion made of worn-out planks of discarded wood. [**Fig. 8.4**] Within this installation of "gritty, dirty, grainy and rough" materials, British photographer Nigel Henderson developed a series of collages composed of images of debris, forming a new take on urban subjectivity.[5] Architectural historian Reyner Banham wrote of the entire installation that "one could not help feeling that this particular garden shed with its rusted bicycle wheels, battered trumpet and other homely junk, had been excavated after an atomic holocaust."[6] The Smithsons's own commitment to debris as late-modern nature continued in their controversial housing project for the Robin Hood Gardens, where the remnants of demolished houses that previously occupied the site were transformed into a new type of collective landscape. [**Fig. 8.5**] Rather than remove the image of debris from the city, the Smithsons locate it as a central aspect of the experience of modern urbanization.[7]

Unlike the British experience in which debris marked the horrific sacrifices of an ultimately victorious state, for the citizens of the defeated nation of Japan, debris lacked any such associations. During World War II, over one million Japanese soldiers and civilians were killed by the Allies, and in one evening 167,171 buildings were destroyed during the firebombing of Tokyo. The bombings of Hiroshima and Nagasaki killed over 200,000 people and imparted a new frightening language of megatonnages and "radio-active" fallout and debris onto modern consciousness. [**Fig. 8.6**] The Japanese architect Kenzo Tange, one of the founders of the Japanese Metabolist movement, wrote one of the more evocative reactions to the horrific and ruinous debris-littered state of Japanese cities after World War II. Reflecting on the destroyed state of Tokyo, he said:

> I cannot dispel from my memory the image of the city (Tokyo) as I saw it immediately after the war at the end of 1945.... The spectacle that met my eyes was desolate.... Around Marunouchi, the financial center of the metropolis, a few of the larger reinforced concrete structures still stood upright... but vast areas of homes, small shops, and stores in lower Tokyo had been razed to the ground.... Here there were not even the mountains of rubble of German towns; the wooden structures had gone up in flames and smoke, leaving the ground covered with black dust and spent embers.... For acres and acres the prospect was one of a grey desert, where every now and then one came across broken crockery, strange green stones (the remains of bottles that had molten because

Fig. 8.6
The remains of Tokyo after firebombing, 1945

Fig. 8.7
The barren debris-ridden landscape of Hiroshima Blast Site: Electric City, by Arata Isozaki, 1968

> of the heat), misshapen sheets of corrugated iron which had barely been covered by some flowering climber that had managed to germinate between one bombing and the next.[8]

Tange contrasts the nature of debris—"the grey desert"—with the more naturalistic form of a plant climbing out of the landscape. But debris and this image of a climbing shoot should be understood as part of one and the same phenomenon—an image that returns Tange's postwar concept of debris back to Le Roy's earliest images of rubble-strewn landscapes. Debris suddenly produces the conditions for a new type of nature—a grey ground that nurtures the most weed-like of verdure.

Tange and other members of the Metabolists negotiated the seemingly impossible nature of reconstruction following such horrible destruction. Many of their projects feature buildings that operate on a new ground, leaving the debris-ridden city as a type of pristine field upon which a new reality is constructed, and some rise directly from images of ruins and debris. Arata Isozaki, one of the youngest members of the group, offered a more frightening engagement with the image and effect of debris-laden worlds. Isozaki's project Hiroshima Blast Site: Electric City contains images of two ambiguous structures rising from the destruction of Hiroshima. The structures

oscillate between appearing to result from the same disaster that befell its surroundings and a new type of building built in relation to this destroyed world. [Fig. 8.7] As in the Smithsons' work, we see an architecture of debris that attempts to imagine the destroyed city as a type of human-produced context—an antinature.[9]

We might think of debris as an undertheorised category within recent architectural culture, simply because the late-modern world has not witnessed a single condensed period of global warfare, but this is not the case. Several contemporary examinations of debris interrogate the architectural imagery of destruction, warfare, and geological transformations first introduced by Le Roy and Viollet-le-Duc. In his utopian project Moonmark, the architectural theorist Jeffrey Kipnis proposes the production of debris as an act of political protest. In this work, he calls for the entire nuclear arsenal on earth to be detonated in one spot on the moon. The resulting explosion would not only scar the lunar surface, but it would eject moon rock into orbit. [Figs. 8.8 + 8.9] Reflecting on his proposed performance of destruction, Kipnis writes:

> The location of the mark was calculated to take maximum advantage of the moon's changing phases, and the orbiting ejected material could have been shepherded with satellites into Saturn-like rings around the moon.... It seemed to provide a fitting testimony to our collective decision to survive and progress beyond our potential for massive self-destruction.[10]

Employing Henri Lefebvre's concept of the production of space as an inherently political act, Kipnis sought a conceptual technique to produce a space through the act of destruction itself. In this, Kipnis's proposal makes debris a tool of projection, and he speaks of a potentially destructive future. Kipnis's techniques share affinities with the strategies of architect Lebbeus Woods, who further explored the appearance of debris, scarring, and the destructive capacity of modern warfare and the geophysical power of the earth. In projects such as Berlin Free Zone, DMZ, and Terrain, Woods examined the production of debris as a type of material for a future architecture. He addresses debris as a form of matter as well as the destructive forces that produce these horrifying wastelands.

Explorations of debris continue in contemporary work. For a public park and buildings in Cologne, Germany, the architects Manuel Herz and Eyal Weizman developed a construction language composed of debris and rubble. In the development of the park's pavilions, Herz and Weizman propose reusing the remains of buildings buried after

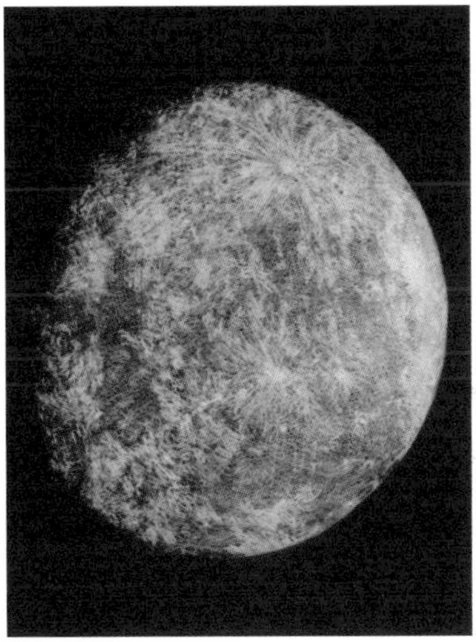

Fig. 8.8
The explosive debris field of Moonmark, by Jeffrey Kipnis, drawing by C. Glenn Eden, 1983

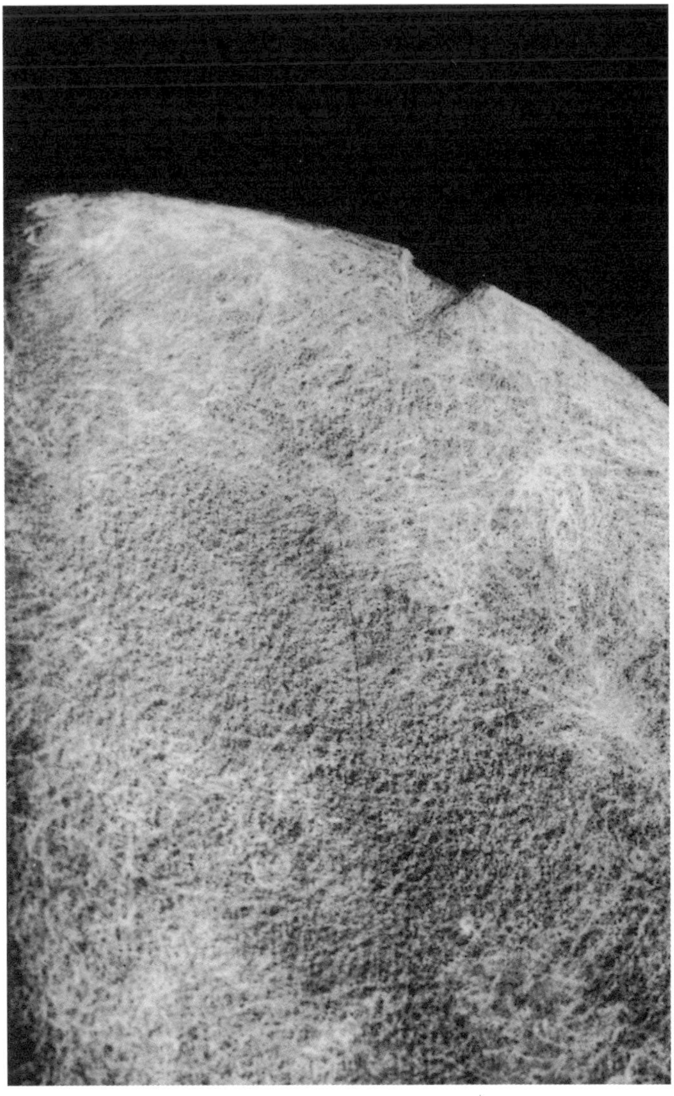

Fig. 8.9
The mark left in the surface of the moon, Moonmark, by Jeffrey Kipnis, drawing by C. Glenn Eden, 1983

the bombing of Cologne during World War II. [Fig. 8.10] Soil, rubble, and debris are heaped over concrete frameworks with elongated windows to extend past the angle of repose of the dumped bits of brick and stone, forming a series of garden pavilions. [Fig. 8.11] The architects propose using shattered matter to build a future out of a violent past; the debris will eventually be planted with sod to become a site for wild flowers, weeds, and plants. Building on the work of the Smithsons, Herz and Weizman transform debris into a true construction system that imbues material qualities that connect with, but ultimately extend past, loss and disaster. [Figs. 8.12–8.14]

From these last two examples we should understand that debris should not become an architectural image of horror, because it cannot be easily reconstituted into its former form. As a construct (or deconstruct), debris is certainly a key component of a new type of subnatural environment born from violence—it is so intimately connected with the destructive capacities of modern production. From its inception, both as a term and a concept, debris refers to cataclysmic social events and the transformation of a building's ground. This, one could argue, has subtle yet important implications: debris, like rubble, returns buildings to their surrounding nature, but unlike theories of ruins and their inherent fantasies of the Picturesque, debris also mutates its surroundings. Because it is often unrecognizable in its original form, and because it often refers to social disasters, debris signifies not only the return of society to nature but it exists as a type of latent hybrid nature in its own right.

Fig. 8.10
Diagrams of construction with debris, Open-Air Parkcafé, by Manuel Herz and Eyal Weizman, Cologne, Germany, 2003

Fig. 8.11 (above)
Site plan, Open-Air Parkcafé, by Manuel Herz and Eyal Weizman, Cologne, Germany, 2003

Fig. 8.12 (right)
Aerial plan, Open-Air Parkcafé, by Manuel Herz and Eyal Weizman, Cologne, Germany, 2003

Figs. 8.13 + 8.14
Views of the proposal, Open-Air Parkcafé, by Manuel Herz and Eyal Weizman, Cologne, Germany, 2003

Dust

1. Teresa Stoppani, "Dust revolutions. Dust, *informe*, architecture (notes for a reading of Dust in Bataille)," *Journal of Architecture* 12, no. 4 (2007): 437-47.

2. John Ruskin, *The Ethics of the Dust* (London: Merrill and Baker, 1877).

3. See the examination of Bataille's concept of dust in Stoppani above and in Yves-Alain Bois and Rosalind Kraus, "A User's Guide to Entropy" *October* 78 (Autumn 1996): 38-88.

4. Joseph Amato, *Dust: A History of the Small and the Invisible* (Berkeley and Los Angeles: University of California Press, 2000), 6.

5. Ibid., 7.

6. Le Corbusier, *Towards a New Architecture*, 56.

7. See a study of this in Marni Reva Kessler, *Sheer Presence: The Veil in Manet's Paris* (Minneapolis: University of Minnesota Press, 2006).

8. See http://www.designboom.com/weblog/cat/9/view/3630/venice-architecture-biennale-08-preview-an-te-liu.html

9. Jorge Otero-Pailos, interview with author, July 2008.

Puddles

1. The literature on Piranesi and Adam is immense. For a summary of relationships between the work of Johann Fischer von Erlach, Piranesi, and Adam, see Robin Middleton and David Watkin, *Neoclassical and 19th Century Architecture*, Vol I & II (New York: Electa/Rizzoli, 1980). For an interesting analysis of the image of the ruin in Piranesi and Adam, see Michael Roth, *Irresistible Decay: Ruins Reclaimed*, eds. C. Lyons and C. Merewhether (Los Angeles: Getty Publications, 1996). For the link between Piranesi and Adam, see A. A. Tait, "Reading the Ruins: Robert Adam and Piranesi in Rome," *Architectural History* 27 (1980): 524-33.

2. On flooding in Paris, see Shelby T. McCloy, "Flood Relief and Control in Eighteenth-Century France," *Journal of Modern History* 13, no. 1 (1941): 1-18; and for additional material on Buache, see Antoine Picon, "Nineteenth-Century Urban Cartography and the Scientific Ideal: The Case of Paris," *Osiris* 18 (2003): 135-49.

3. Antoine Picon, *French Architects and Engineers in the Age of Enlightenment* (Cambridge, UK: Cambridge University Press, 1992).

4. Panayotis Tournikiotis, *The Historiography of Modern Architecture* (Cambridge, MA: MIT Press, 1999), 103.

5. From the architects' unpublished project boards.

6. Philippe Rahm, *Mollier Houses*, exhibition board for Anxious Climate, 2006.

Mud

1. Chambers was actually a Swedish-born, French-trained architect noted for his architectural experimentation. In England, he pushed accepted architectural concepts beyond Palladianism.

2. Adrian Forty, "Primitive, the Word and Concept," in *Primitive: Original Matters in Architecture*, ed. Jo Odgers, Flora Samuel, and Adam Sharr (New York: Routledge, 2006), 3-14.

3. Marcus Pollio Vitruvius, *Vitruvius: The Ten Books of Architecture*, trans. Morris Hicky Morgan (New York: Dover, 1960), 39-40.

4. John Harris and Michael Snodin, eds., *Sir William Chambers: Architect to George III*, (New Haven: Yale University Press, 1996), 70-71.

5. On the mud of the Thames, see Cohen and Johnson, *Filth*.

6. On the Embankment, see Dale H. Porter, *The Thames Embankment: Environment, Technology, and Society in Victorian London* (Akron, OH: University of Akron Press, 1998); and Lynda Nead, *Victorian Babylon: People, Streets, and*

Images in Nineteenth-Century London (New Haven: Yale University Press, 2000), 53-56.

7. Nead, *Victorian Babylon*, 56.

8. See the work of Warwick Anderson for an interesting discussion of colonial engineering projects.

9. Hassan Fathy, *Architecture for the Poor: An Experiment in Rural Egypt* (Chicago: University of Chicago Press, 1973), 4.

10. Many of the Egyptian poor for whom Fathy designed mud housing felt alienated by his use of domes—a form more typically associated with religious buildings than residential buildings.

11. See Raymund Ryan, *N3: The Irish Pavilion at the Seventh International Biennale, Venice* (Dublin: Haus Publishing, 2001).

12. Seamus Heaney, "Kinship," in *North* (London: Faber & Faber, 1975), 41.

Debris

1. On rubble, see Jeff Byles, *Rubble: Unearthing the History of Demolition* (New York: Harmony Books, 2005).

2. See Robin Middleton's introduction in Julien-David Le Roy, *The Ruins of the Most Beautiful Monuments of Greece* (1753; Los Angeles: Getty Publications, 2004).

3. Eugène Emmanuel Viollet-le-Duc, *Annals of a Fortress: Twenty-Two Centuries of Siege Warfare*, trans. Benjamin Bucknall (New York: Dover, 2007).

4. Ludwig Hilberseimer, *The Nature of Cities: Origin, Growth, and Decline, Pattern and Form, Planning Problems* (Chicago: Paul Theobald, 1955).

5. One of the best histories of the installation can be found in Sarah Williams Goldhagen, "Freedom's Domiciles: Three Projects by Alison and Peter Smithson," in *Anxious Modernisms: Experimentation in Postwar Architectural Culture*, ed Sarah Williams Goldhagen and Réjean Legault (Cambridge, MA: MIT Press, 2000); the quote is from page 84.

6. Reyner Banham, *The New Brutalism: Ethic or Aesthetic?* (London: Architectural P., 1966), 85.

7. On the Smithsons's early work, see *The Independent Group: Postwar Britain and the Aesthetics of Plenty*, ed. David Robbins (Cambridge, MA: MIT Press, 1990); Goldhagen, "Freedom's Domiciles"; and Peter and Alison Smithson, *The Charged Void: Architecture* (New York: Monacelli Press, 2001).

8. Paolo Riani, *Kenzo Tange* (New York: Hamlyn, 1970), 7-8.

9. On the Metabolist movement's response to post-World War II Japan, see Cherie Wendelken, "Putting Metabolism Back in Place: The Making of a Radically Decontextualized Architecture in Japan," in *Anxious Modernisms: Experimentation in Postwar Architectural Culture*, ed. Sarah Williams Goldhagen and Réjean Legault (Cambridge, MA: MIT Press, 2000).

10. Kipnis, Jeffrey, "Moonmark," *Assemblage* 16 (1991): 11.

LIFE

Weeds

Insects

Pigeons

Crowds

Weeds

We have examined subnatural atmospheres (and how architects view them as potentially confrontational to the existence of modern urban society) and subnatural forms of matter (excavating the ways that architects imagine them as markers of social transformation, history, and time), but in the final four chapters, we focus on subnatural forms of life. As in the previous eight chapters, we will examine how life-based subnatures are simultaneously marginalized and embraced in architectural discourse. Unlike previous sections, the chapters that follow consider how subnatural forms of life become metaphors of new forms of subjectivity in modern society. Equally significant, subnatural life often contains a strong provisionality. No form of life is inherently subnatural; rather, relative to architecture, life becomes subnatural when it makes us question the dominant social role of architecture. The provisional and relative notions of subnatural life are comparable to the concept of dirt undertaken by cultural anthropologist Mary Douglas. In the book *Purity and Danger*, Douglas asserts that things, people, and practices become dirty when they are "matter out of place." Nothing is inherently dirty or dirtlike; rather, dirt is a social category that we assign to specific types of social relations. For Douglas, dirt lacks any fundamental physical quality. Instead, it is a relationship. Think of the undesirability of food that falls on the floor; we do not want to eat food from the floor, because the floor is "dirty" no matter how many times one sweeps or mops it. If dirt is "matter out of place," then we might paraphrase Douglas's study for this chapter to say that weeds are plants out of place. Weeds are not inherently unwanted, useless, or invasive. In fact, of all the forms of lesser nature reviewed in this book, weeds are the hardest to define, because botanists cannot agree on what exactly differentiates a weed from other nonweed species. The only constant between weeds and other plants is their socially determined undesirability. For example, dandelions might be one of the largest nuisances of contemporary American lawns, but it is also surprising to know that farmers in the nineteenth century had difficulty growing dandelions as an edible plant for their greens.[1] Plants become weeds when they are out of place in agricultural settings, but they also become weeds in other non-natural settings when they disrupt an inherent order. For example, weeds are those plants that get in the way of the programs, agendas, or desires that we project into spatial constructs. Ivy quickly becomes an invasive species when it disrupts the functions of windows or walls.

Over the course of architectural thought, various individuals have embraced the seemingly undesirable weed as a thing and metaphor for repositioning architecture. Often, they are used as metaphors and as literal matter by architectural thinkers seeking to transgress socially accepted boundaries. Where there is a new movement, there often is imagery of weeds, whether with respect to Picturesque landscapes, organicist architectural theory, or poststructuralist approaches. Many of these movements contain such images. Thus, not only do weeds get in the way of space, but they are perceived as alien encounters, suggesting new roles that spatial formations might yet take.

Consider some of the earliest images of plants in architectural discourse; they are, in some sense, weeds. Vitruvius wrote of an ancient myth that attributes the origin of the Corinthian column to the intrusion of a weedike plant on a funerary urn:

> A freeborn maiden of Corinth, just of marriageable age, was attacked by an illness and passed way. After her burial, her nurse, collecting a few little things which used to give the girl pleasure while she was alive, put them in a basket, carried it to the tomb, and laid it on top thereof, covering it with a roof-tile.... This basket happened to be placed just above the root of an acanthus. The acanthus root pressed down meanwhile though it was by the weight, when springtime came round put forth leaves and stalks in the middle, and the stalks, growing up along the sides of the basket, and pressed out by the corners of the tile through the compulsion of its weight, were forced to bend into volutes at the outer edges. Just then Callimachus... passed by this tomb and observed the basket with the tender young leaves growing round it. Delighted with the novel style and form, he built some columns after that pattern for the Corinthians, determined their symmetrical proportions, and established from that time forth the rules to be followed in finished works of the Corinthian order.[2]

In this myth, Callimachus draws an acanthus; the Greeks used the term *acanthus* for any number of varieties of thistle-leaved plants. Some were of the kind we would consider weeds, with sticky unctuous leaves and thorny stems. Others were venerated for their use in Greek and Roman gardens. The story of Callimachus and the acanthus was illustrated by architectural theorists from the fifteenth through the eighteenth centuries in ways both weedlike and not. In a premodern depiction of the scene by the Italian architect Antonio Rusconi, the acanthus and urn are shown in isolation as a uniform composition, but in later renderings the acanthus appears

Fig. 9.1
Callimachus devises the Corinthian capital, etching by Roland Fréart de Chambray, 1650

as a vibrant, almost invasive element. In the French architect Roland Fréart de Chambray's early-modern illustration of the Corinthian story (1650), the acanthus invades a classical scene consisting of a pyramidical funerary monument and the urn described by Vitruvius. [**Fig. 9.1**] Behind the urn and the acanthus, another weed climbs the pyramid-shaped building, and nature appears as a transgressive animated force against buildings. In Fréart de Chambray's etching, weeds begin to challenge architecture while clearly providing artistic inspiration.

Such socionatural tensions were explored more deeply within the Picturesque aesthetics of the late eighteenth and early nineteenth centuries, when the unexpected appearance of plants in general, and weeds more specifically, challenged classical and neoclassical conceptions of beauty and other popular forms of the eighteenth-century landscape. Theorists of the Picturesque in landscape and architecture advanced a more unshackled form of nature, represented by roaring streams, overgrown pastures, wild flowers, and weeds. In the poem *The Landscape* (1794), by the preeminent theorist of the Picturesque, Richard Payne Knight, wrote:

Fig. 9.2
A Picturesque landscape garden by Thomas Hearne, 1795

Let clustering ivy o'er its sides be spread,
And moss and weeds grow scatter'd o'er its head.
The stately arch, high-rais'd with massive stone;
The ponderous flag, that forms a bridge alone;
The prostrate tree, or rudely propt-up beam,
That leads the path across the foaming stream;
May each the scene with diff'rent beauty grace,
If shewn with judgement in its proper place.[3]

Knight commissioned the artist Thomas Hearne to illustrate the overgrown settings, rude architecture, and nature described in *The Landscape*, differentiating this rough image against more popular forms of English landscape design. In one of these commissioned illustrations of a non-Picturesque garden design, nature is arranged into a more manicured composition, centered on tree groves and a winding brook; the scene establishes multiple expansive views. In the Picturesque landscape, weeds line a craggy brook replete with fallen trees, having an insular visual effect. [**Fig. 9.2**] In the latter image, weeds represent not only something wild but something "unimproved," left as it was. In a sense, weeds represent a type of nature that has not been transformed by industrial agricultural development. At this time in Britain, fields were undergoing significant transformation and modernization in efforts to make agriculture a more profitable business enterprise. Weeds, as useless plants, came to signify untransformed landscapes having no cash value.[4]

By the nineteenth century, weeds emerged as an inherently confrontational force against established forms of social order. Such a concept can be traced to the German aesthetic philosopher Johann Wolfgang von Goethe, who examined plants as conveyors of a vitalist form that was expressive of life's inherent energy. In his exploration of plants, Goethe imagined the existence of an archetypal yet fictional plant form from which all other plants could be derived.

In "discovering" this archetypal plant form, Goethe believed that he could further identify the formal principle behind any number of living things. He believed the study of such forces in nature might lead to the knowledge of a specific, natural formal principle, and that such a living spirit might impact the realization of any variety of social forms, particularly art. [Fig. 9.3] Goethe's theories of natural vitalism influenced Romantic, Transcendental, and Organic literary and architectural theory.[5] Within this writing, weeds exemplified those uncultivated life forms lurking beneath the natural world, opening up paths to an alternative modernity rooted in the revolutionary potential of free nature. Several examples illustrate these interlacing concepts. For example, the nineteenth-century American writer Henry David Thoreau saw the weed as something having an inherent organization that was under threat by civilization. He believed that we have no right to destroy "their delicate organization so ruthelessly" or to create "invidious distinctions" between weeds and the more socially desirable plants that might be removed from gardens. Similarly, Nathaniel Hawthorne wrote that weeds were "more essential to the well-being of the world than the most precious fruit or grain."[6] Within architecture, the Transcendentalist Louis Sullivan wrote often of the weed as an unlikely source of inspiration. In the passage "Letter to My Uncle," Sullivan wrote a poem to a weed:

> I made a little one to a weed the other day. I like weeds: they have so much "style" to them and when I find them where they have grown free they seem most interesting and suggestive to me. I think I'm something of a weed myself.... And then there are so many of them, and they differ so much in shape, color and arrangement; the form follows the function so beautifully as you would say. I don't know the names of any of them—being city-bred.... I wish I knew the names of the little rascals; then it seems to me, I could talk to them better.[7]

And in a later work he wrote more broadly of the weed as a program for architecture:

> We allude to the weed as common, as wild. It may be "common" as we say, it may be "wild" as we say, but it is a miracle nevertheless. It is a most impressive symbol.... The weed is of the work of Nature, remains with Nature and is, therefore, "natural" as we say. The man, strangely enough it might seem, has departed from Nature whence he came. Strangely, he has, in pride of intellect, denied instinct: called it a weed, and wild, as it were.[8]

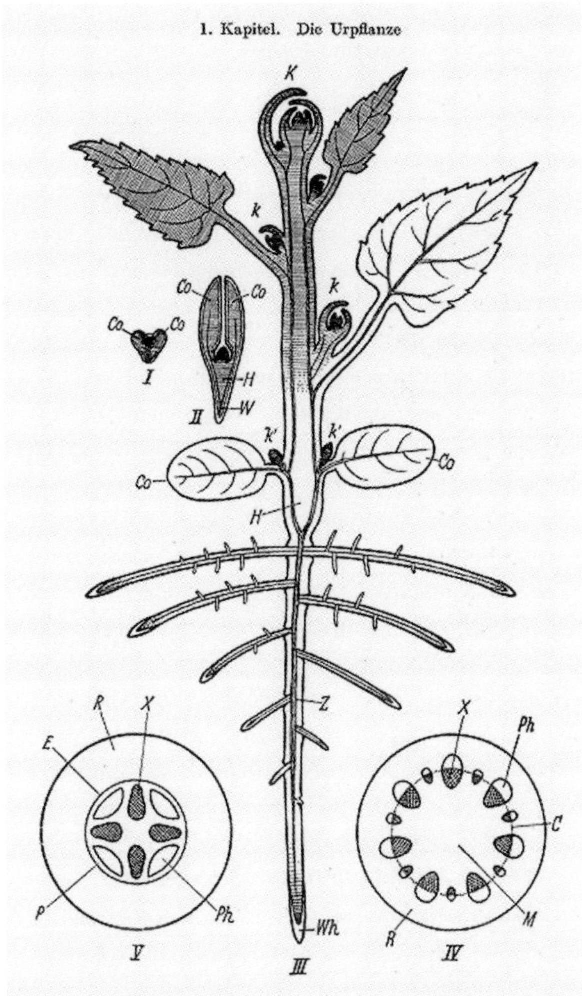

Fig. 9.3
The Ur Plant, drawing by Johann Wolfgang von Goethe, mid-19th century

Sullivan is an unexpected theorist of the weed. As a designer of ornament, he reinterpreted classical and Romanesque acanthus motifs in more weedlike plant forms and developed new designs based on weeds. For the Auditorium Building (1890), he reconceptualized an acanthus ornamentation developed by the American architect H. H. Richardson as a "spiky-edge thistle leaf," a common weed of the American plains.[9] [**Fig. 9.4**] In later projects, Sullivan's ornamentation simultaneously invaded and emphasized the formal concepts of his tall building designs, as well as those of Dankmar Adler. Weeds are unleashed to suggest the furnishing of new spatial forms within the American city.

Sullivan's apprentice Frank Lloyd Wright also wrote of weeds, undoubtedly inspired by the concepts of his employer, but unlike Sullivan, Wright embraced the weed both as metaphor and as possible matter in architecture. In the section titled, "The Hoe," a protracted diatribe on weeding from his *Autobiography* (1943), Wright wrote:

> The wielder of the hoe would wonder why weeds couldn't be studied, possibilities found and then maybe cultivated. The "crop" eliminated.... Tobacco was a weed once.... And tomatoes were once thought by Europeans to be poison.... Nearly everything was a weed once upon a time.... What vitality these weeds had! Pulsey, for instance, Chess (velvet weed). Pigweed.... Would the weeds become feeble, if they were cultivated and "crops" become as vigorous as "weeds" if able to flourish on their own? What of such science and art?

Wright expands the weed beyond Sullivan's vitalism of form into an entire social program that emphasizes vigor and independence. In the late nineteenth century, Wright designed vases and planters specifically for the display of weeds in his Chicago residential commissions. Drawing on Romantic and Transcendentalist disdain for cultivated flowers, Wright presents a place for weeds within the American home. His exploration of the weed contains enormous novelty, yet the incorporation of displays of weeds in his house designs seems ironic in light of his own and Sullivan's writings. Within Wright's work we see a domestication of the wild and uncultivated aspects of weeds, even as these plants are called on as metaphors of social and mental transformation.

Weed metaphors extend through postwar American explorations of nature and space. In the 1970s, the Chicago-born American architect and writer James Wines explored a more populist yet seemingly transgressive position for weeds and other wild nature in a series of projects for the United States–based Best Products company; entire suburban big-box retail stores were outfitted with weeds and scrub. The larger social program of the weed and the transformative power of nature is dropped in favor of a more obvious play on ironic attitudes toward suburban forms of nature. Weeds and nature in general become an invasive form of entertainment—a kind of anti-aesthetic commentary on suburban leisure and labor. Weeds also enjoyed resurgence as powerful metaphors in contemporary American architectural theory and practice. Inspired by the earlier writings of Goethe, as well as by Gilles Deleuze and Félix Guattari's writings on the weedlike rhizome, American architects sought ways to apply the weed metaphor to a new type of form-driven practice infiltrating the spaces of late-modern American cities.[10] An early and exemplary project that employed Deleuze and Guattari's metaphor is the

Fig. 9.4
***Impromptu*, drawing, by Louis H. Sullivan, 1924**

Stranded Sears Tower (1991) by American architect Greg Lynn. The tower appears to ensnarl its surroundings in a new type of animistic space. The project is weedlike in that it invades the streetscape and appears to exist outside any known form of urban zoning that might otherwise temper its form. It is a weedlike counterimage to the existing Sears Tower. [**Fig. 9.5**]

These recent evocations of the weed by Wines and Lynn suggest a type of alterity that might be found by using marginalized plants and referring to their forms. Yet, for all the recent efforts to capture the weed as matter and metaphor, many late-modern engagements with the weed in architecture seem quite tame. And this might be the ultimate irony of employing the weed: as something defined in relational terms only, as soon as the weed is put to use—conceptually, metaphorically, materially—it no longer remains a weed. Ultimately, what is called for is an architecture that produces the weed as an active concept in real-existing space; that is, a building that transforms plant matter not previously considered a "weed" into just that. This approach, which we might call with some irony *weeding*—a technique that lends a weedlike character to the nature adhered to architecture—can be found in recent work by R&Sie(n) Architects, West 8, and Cero9. These architects employ plant matter as an element that appears to potentially consume its surroundings. Of these, the Magic Mountain by Cero9, a type of vegetal covering for an unsightly power plant in Ames, Iowa, exemplifies the new approach. For this manmade verdant construct, Cero9 proposes:

> To totally cover with a membrane of roses, lights and honeysuckle the fragmented volumes of the power station. The membrane creeps above the highest parts of the power station and transforms the building into a vertical garden with living technified walls. The species chosen for the construction of this vegetation crust are a combination of rosebush creepers against a green background of honeysuckle... we propose to use gardeners' ancestral techniques of genetic selection to induce a modern image in the uncomfortable presence of the power station.[11]

They envision this tower of plants as not only taking over the form of the power plant but as scattering the petals and pollen of it upon the building's surrounding area. The building takes plant matter that we might not necessarily consider a weed and turns it into something invasive, something weedlike. [**Figs. 9.6–9.10**] A similar approach is found in R&Sie(n)'s Lausanne City Museum, the nMBA. In this project, R&Sie(n) reiterates the conceit of Lynn; they design a building as a formal analog to the weed's biological desire to invade and spread. Yet they also consider the way plant matter potentially moves around the entirety of the structure, overwhelming the architectural surfaces of the museum. [**Figs. 9.11–9.14**] In an evocative text, R&Sie(n) writes:

> The nMBA is a place of illusions; it exists where "the wild," "weeds," "urbanized," and "artificial" nature converge.... Interlaced like a rhizome, progressing like a bed of coral and entangled like stick insects, it forms a phasmida.... More a landscape than an urbanism; more a forest than architecture. A project that plays with its natures. "Weeds" that become the local woods that are then populated with animals, like an amphibian world that has been emancipated from water, having appeared freely and spontaneously.[12]

The nMBA is designed to both facilitate the production of weedlike plants and to perform in ways analogous to the weed. In many ways, R&Sie(n) extends the weed metaphor of Lynn's earlier explorations in Chicago. In considering the spreading and consuming analogies of weeds, a recent project by West 8 transforms potted flowers into something containing weedlike resonances. [**Figs. 9.15 + 9.16**] For a landscape to memorialize the bombing of Rotterdam, Adriaan

Geuze of West 8 developed an image of flames, out of flowers, appearing to consume its surroundings:

> Rotterdam's inner city was bombed on May 14. The Schouwburgplein is situated in the heart of Rotterdam destroyed by the bombing and the ensuing fires that broke out. City on Fire/City in Bloom makes the flames symbolically visible in a blaze of flowers. These flowers are a reminder of the blossoming of Rotterdam in the aftermath of the war.[13]

In these works, Cero9, R&Sie(n), and West 8 embrace the relativist notion of plants as colonizers of space, as unwanted, as matter out of place, as "the weed," and they discover ways to use architecture to bring plants into realms, positions, and forms in which they may not belong.

Fig. 9.5
Stranded Sears Tower, by Greg Lynn/Form, 1991

Figs. 9.6
Model, Magic Mountain, by AMID/Cero9 Architects, Ames, Iowa, 2004

Figs. 9.7 + 9.8
(opposite) Renderings, Magic Mountain, by AMID/Cero9 Architects, Ames, Iowa, 2004

PHILLIPS
66
HOME FURNITURE
35
69
ATM
CARWASH

2 The structural flowerpots are attached to the existing walls of the industrial installation, providing a corridor between the plants and the walls for eventual maintenance.

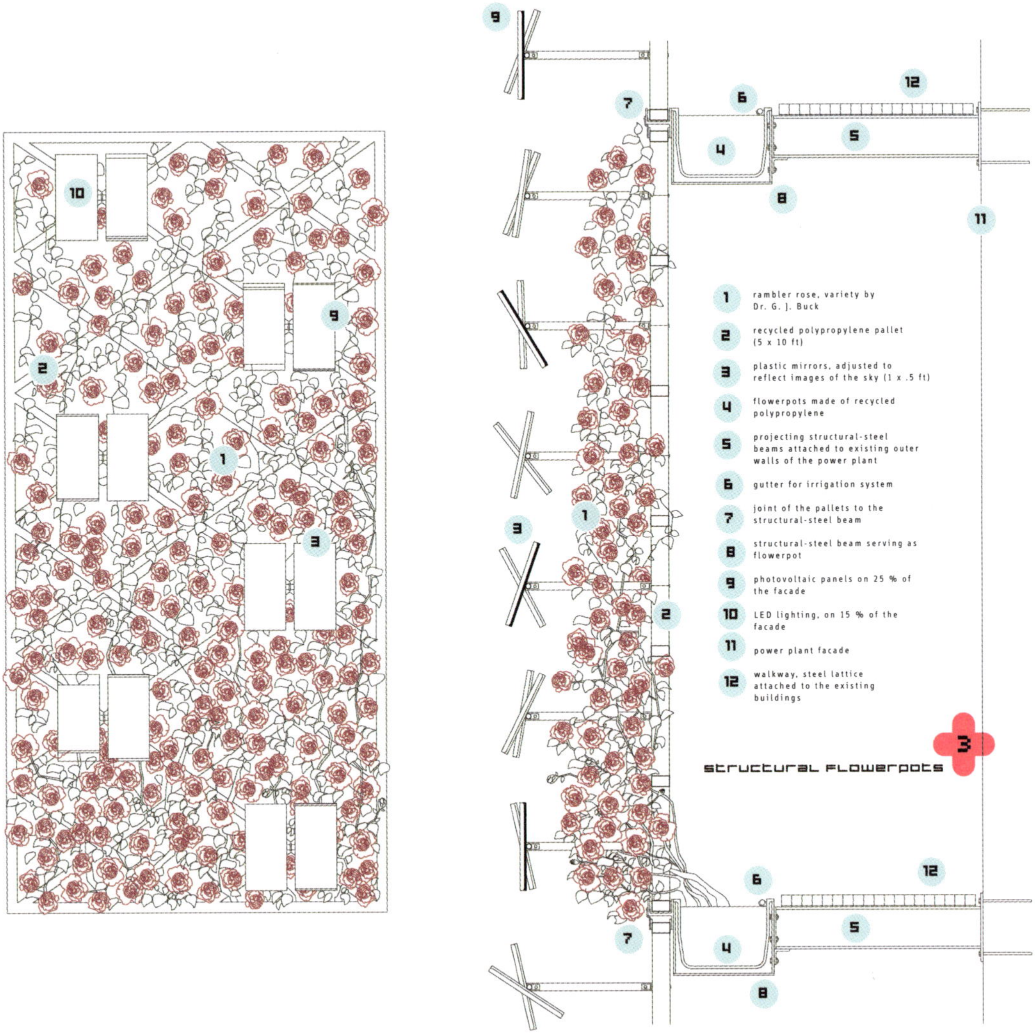

Fig. 9.9
Construction system, Magic Mountain, by AMID/Cero9 Architects, Ames, Iowa, 2004

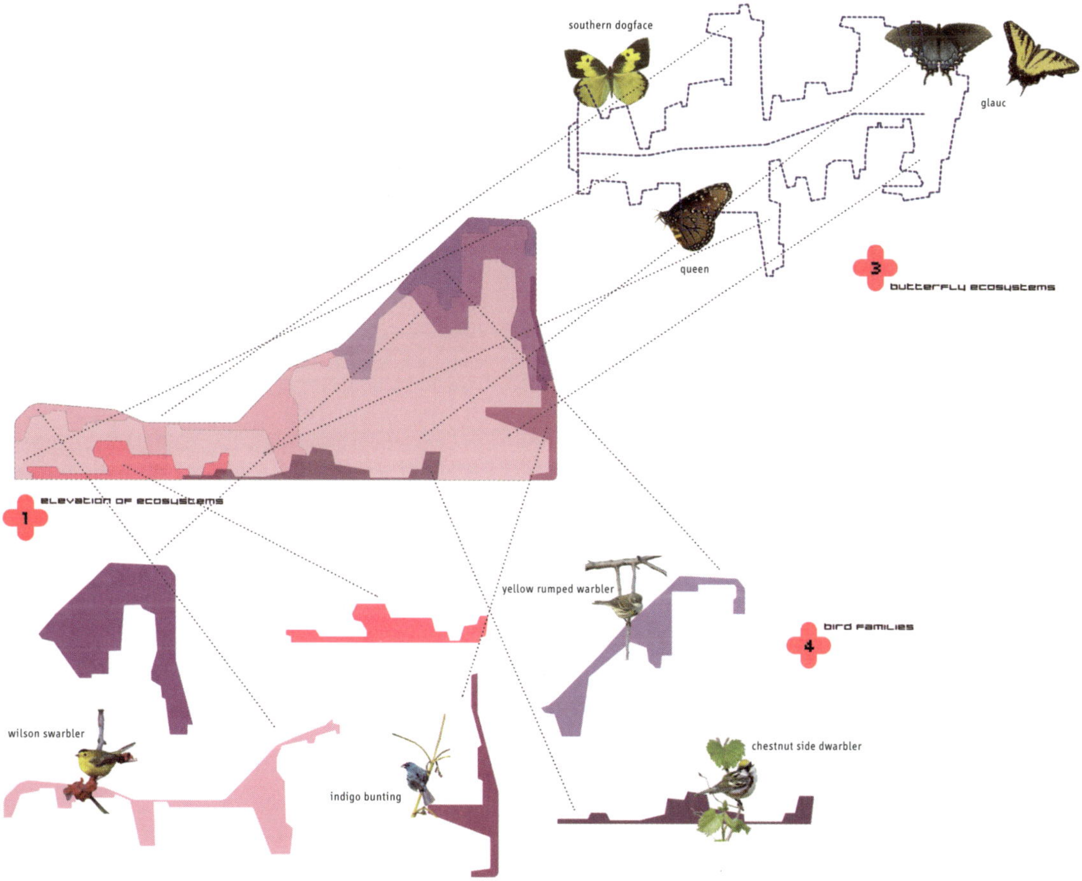

Fig. 9.10
Ecosystem diagram, Magic Mountain, by AMID/Cero9 Architects, Ames, Iowa, 2004

Fig. 9.11
(above) Aerial rendering, nMBA, by R&Sie(n) Architects, Lausanne, Switzerland, 2005

Fig. 9.12
(below) Perspective view, nMBA, by R&Sie(n) Architects, Lausanne, Switzerland, 2005

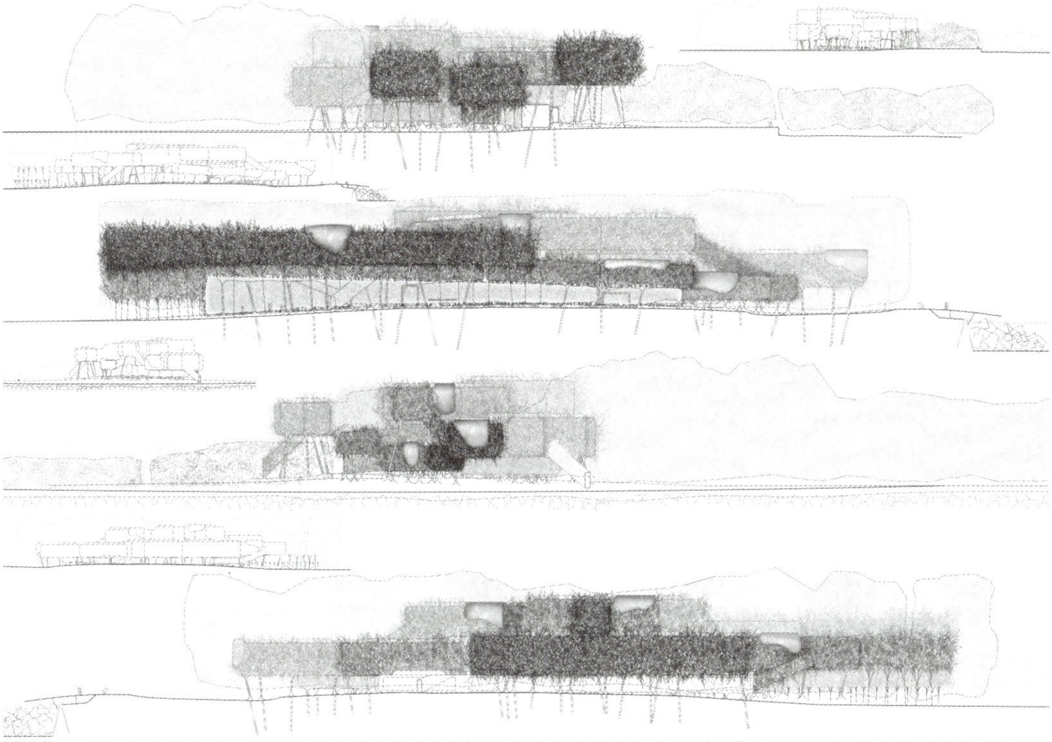

Fig. 9.13
Elevations, nMBA,
by R&Sie(n) Architects,
Lausanne, Switzerland, 2005

Fig. 9.14
Detail of wall system, nMBA, by R&Sie(n) Architects, Lausanne, Switzerland, 2005

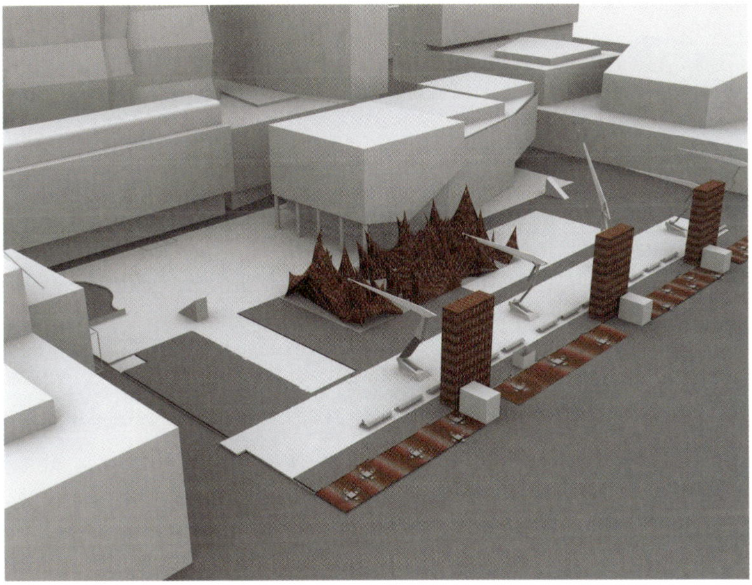

Figs. 9.15 + 9.16
City on Fire/City in Bloom, by West 8, Rotterdam, the Netherlands, 2007

Insects

It is difficult to imagine that architects might reconsider the relationships, both conceptual and actual, between buildings and insects. Insects are anathema to our concept of the architectural interior or any notion of architecture as refuge. They infest architecture, often in repulsive ways; some insects, like termites, actually damage the structures of buildings. In cities, cockroaches are vilified as frightening nuisances and carriers of disease. Historian Ben Campkin has explored how in mid-twentieth-century London, working class protesters paraded in front of their insect-infested housing with effigies of roaches. They burned these effigies in public protest to express disgust at their living conditions, but as truly unsettling as insects are, like so many of the forms of subnature explored in this book, architecture has a more complex relationship to insects than our immediate reactions to them would suggest. Insects are a common metaphor for various forms of architectural production, and as this short chapter will reveal, insects have served as catalysts for rethinking the experiences of human beings in modern spaces.

Insects can swarm, like gnats in a field; they can gravitate, like flies; they can be drawn to particular occurrences, like moths to the flame; and they can build structures, like bees build their hives. The particular movements and forms of congregations of insects have often served as inspiration for designing forms within architecture, and some social structures among insects have been valued as key architectural metaphors across several practices and timeframes. According to architectural historian Juan Antonio Ramírez, the *beehive metaphor*, as he terms it, has served as a potent conceptual bridge between insects, society, and architecture.[1] The beehive metaphor is literally the invocation of structures within beehives, particularly the hexagon matrix of the honeycomb, within architectural design. According to Ramírez, such structures can be seen in practices as diverse as Antonio Gaudí, Peter Behrens, and Frank Lloyd Wright. These architects' and others' use of the beehive metaphor extends and transforms the use of the beehive by earlier social commentators as the symbol of organizational power for harmonious societies. [**Fig. 10.1**] For example, during the French Revolution the beehive became a symbol of the new republican virtues of liberty, freedom, and fraternity; an image of revolution was often shown alongside those of beehives. The hive represented a collective industriousness. A somewhat more troubling use of the beehive as a governmental metaphor appeared in Victorian England. An image of the "British Beehive" depicts the interacting structure and hierarchy

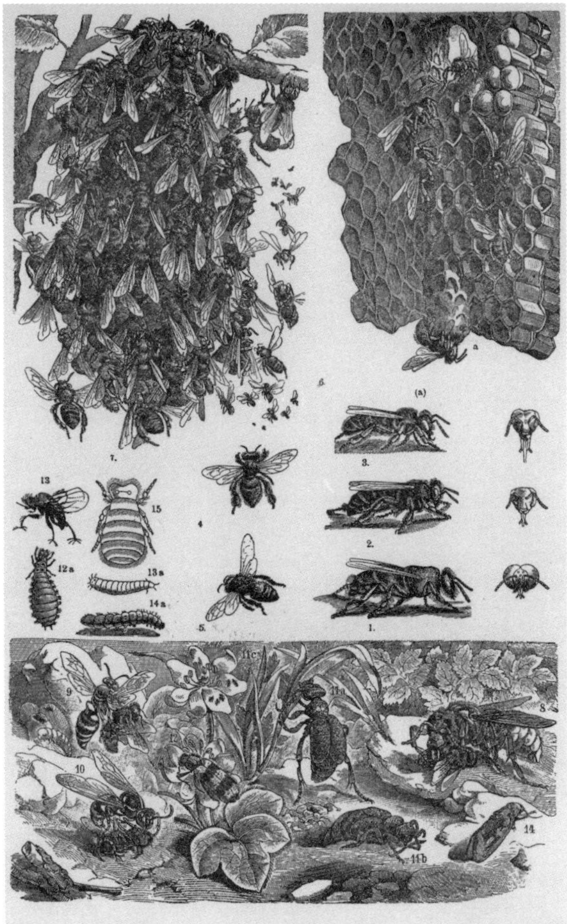

Fig. 10.1
Bees creating a hive, mid-19th century

Fig. 10.2
The British Beehive, 1867

of the political economy of mid-nineteenth-century imperial England. [**Fig. 10.2**] Cleaners populate the lowest parts of the hive, with financiers and ministers at the top—all encircling the queen at the center. The actual structure of the beehive is modified to produce a more absolute and strict hierarchy centered on the hive's queen.

Architects such as Gaudí, Behrens, and Wright appropriated the beehive metaphor, reworking its social associations into spatial concepts. Wright, for example, in his Hanna House in Palo Alto, California, and Arizona State Capitol proposal used the beehive metaphor as a new type of planimetric organizational principle to counter the banality of "gridded environments." Where the grid represented a juncture between structural science and cognitive spatial perception, Wright employed the hexagonal unit as the structure of space. In place of the grid's Cartesianism, Wright drew on imagery from the insect world to project a more naturalistic (he would call it organic) link between the organization of space and forms of socialization.

Fig. 10.3
Rendering of the Arizona State Capitol, by Frank Lloyd Wright, 1956

Fig. 10.4
Installation detail of honeycomb structure, Manifold, by Matsys/ Andrew Kudless, 2004

[Fig. 10.3] Unlike the image of the English hive, however, Wright's construct lacks a central space of domination. He does not use the hive metaphor to intensify the power of a particular individual; rather, he reworks the hive into a landscape-driven concept. Such hive imagery continues throughout late-modern architectural practices, particularly in the development of structural systems invoking the hexagonal matrix. [Fig. 10.4]

If the structures produced by insects could provide metaphors for architectural production, the behaviors of insects and their surrounding settings also provided architectural inspiration. Where the bee metaphor might serve as a model of industrious construction by an insect, the parasite metaphor—often invoking the flee, mosquito, tick, or wasp—was inspired by the interactions of insects with other biological entities. These metaphors, which began to appear in the late-1970s and mid-1980s, paralleled an increased interest in the architect's role as agitator of an accepted social structure. The parasite metaphor offered an image of the architect either in direct confrontation with the spaces that constitute contemporary social life or in a position for reworking them from within. In his exhibition Deconstructivist Architecture, architectural historian Mark Wigley turned to the parasite as a new image for the formal development of architecture. Wigley conveyed some of the above aspects of the parasite metaphor when he wrote of a building addition by the firm Coop Himmelb(l)au, describing it as "a subversive alien, a foreign body that already inhabits the interior and cannot be expelled without destroying its host."[2] The metaphor emphasized a tension between the world as it exists and a new architecture directed against it. Yet, unlike modernist architectural tropes in which buildings rise above the existing city, the parasitic building infests the existing city. For Greg Lynn, the parasite metaphor offered less of a confrontational role for architectural form and more of a cunning reworking of form and context. In a critique of Wigley's use of the parasite metaphor, Lynn wrote that the "parasite has to configure the possibility for its own existence."[3] That is, the parasite produces stability between itself and its host. Turning again to a direct insect metaphor, Lynn discussed the interchanges between digger wasps and a species of orchid that mimics the female wasp's form. Lynn sees this as a new codependent form of parasitism. By mimicking the form of the female wasp, the orchid essentially uses the wasp to help pollinate its flowers. Lynn called this collective wasp-orchid a "fusional multiplicity," and he sought to arrive at a formal architectural equivalent of the dynamics of this insect-plant exchange.[4] Such insect metaphors continue in contemporary practice, emphasizing the idea of the insect's swarm or its ways of organizing its world to effect transformations in climate. The former investigation emerges from the constellation of ideas known as field theory, which considers the laws governing the organization of unstable matter, and the latter emerges from environmentalist engineering, which in this instance examines how insects modify the environments of their hives.[5]

In addition to insects' forms of socialization, architects and the theorists that have inspired them have also explored the subjectivity of insects. Insect subjectivity could inform a new type of experience in which the world of modern human societies is reconsidered through an imagined insect physicality and consciousness. Architects draw on themes explored much earlier in literature. According to the historian Cristopher Hollingsworth, the notion of human-insect subjectivity—what he labels the "insect as self"—appears most radically within Franz Kafka's early-twentieth-century short story *The Metamorphosis* (1915).[6] In this story, the protagonist Gregor Samsa awakens to find himself transformed into an enormous insect—translated variously as an enormous "vermin" or "cockroach." Samsa's new verminous self hides in his room, alienated from the staff of the inn where he stays and where his employers seek him out. Through this insect creation, Kafka explores the limits of the social structure of middle-class Viennese society and the boundaries of human socialization.

This literary conceit was explored in a more philosophical framework in the 1970s and 1980s by the philosophers Gilles Deleuze and Félix Guattari and in a somewhat modified form by the anthropologist Donna Haraway.[7] They all seized the implications of the human-insect transformation as a type of political strategy that enabled one to operate as a radical new type of nonhuman subject. Deleuze and Guattari drew on *The Metamorphosis* directly in their concept of "animalization" or "becoming-animal." Through this concept, they imagined that by "becoming-animal" the human subject could position itself against categories such as family, home, and state. The animal, as Kafka's story demonstrates and as these philosophers further articulate, cannot accede to these formulations. Haraway explored, more specifically, the political position an animalization might engender through her concept of the cyborg—a hybrid of animal, human, and machine. In illustrating this, she seized upon a description of a protest against nuclear testing in the 1970s, in which a group of women took on the character of a desert worm through the creation of a costumed performance. They writhed inside their costume, attempting to develop a hybrid human-insect identity as an expression of political position. Building on and mirroring many of these modernist notions of animalization, architects projected performances and building designs that suggested new rapprochements between the worlds of insects and humans. Insect imagery moves throughout the constellation of practices and concepts known as experimental architecture that appeared from the late-1960s to the mid-1970s. These include well-known concepts, such as Ron Herron's Walking City, and other more obscure realizations of the insect subject in architecture. One of the more subtle and less literal of these was Coop Himmelb(l)au's *Heart City: The White Suit* (1967). [**Fig. 10.5**] This early project consisted of a helmet that relayed imagery and odors, and a pneumatic vest that applied various pressures to its wearer. The project explored a technologically driven architecture of media and sensation, but it resorted to insect imagery to give the exploration visual character. Linking technological, sensorial, and insectoid experiences and images, the project intersects with the cyborg realities imagined by Haraway and her predecessors.

In contemporary practice, the insect image continues to be a robust aspect of experimental work. In the 1980s and 1990s, it

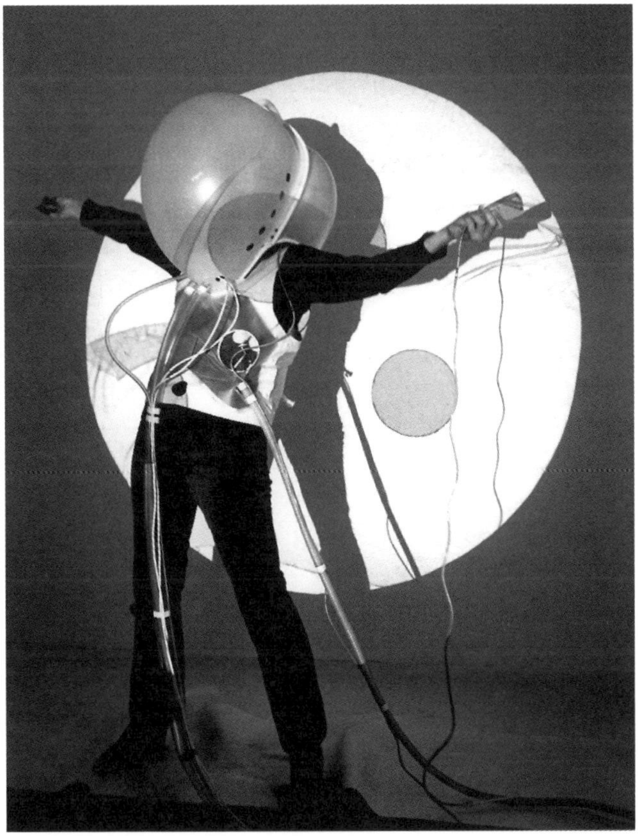

Fig. 10.5
Heart City: The White Suit,
by Coop Himmelb(l)au, Vienna, Austria, 1967

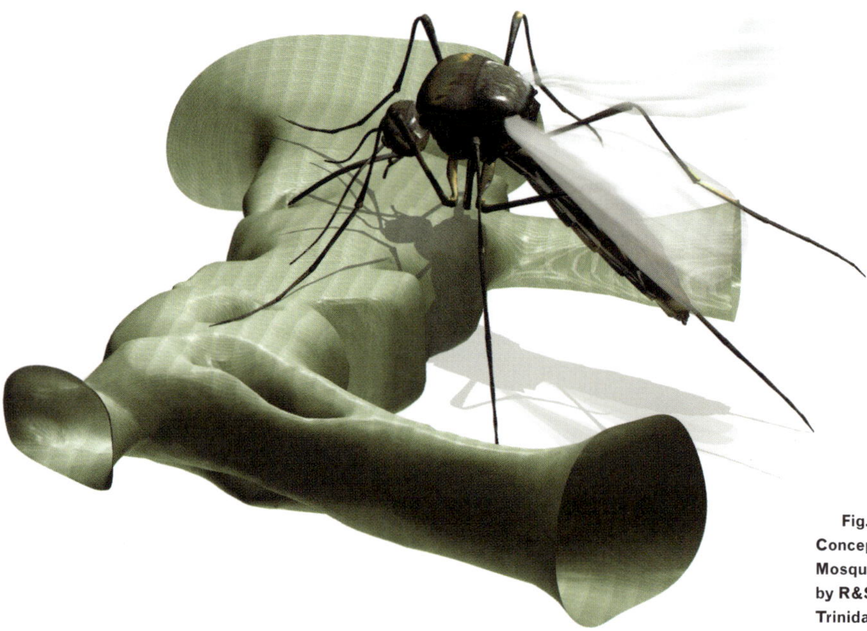

Fig. 10.6
Concept rendering of the Mosquito Bottleneck, by R&Sie(n) Architects, Trinidad, 2003

continued in the work of Coop Himmelb(l)au and entered some of the images produced by Lebbeus Woods and Michael Sorkin. We might argue that the insect metaphor rivals the "machine" metaphor as the more obvious avant-garde architectural image. Most recently, the twin concepts of insect organization and insect-as-self conjoin in the contemporary project Mosquito Bottleneck by R&Sie(n) Architects. [Fig. 10.6] In this project, the architects explored the development of a house for a mosquito-infested area of Trinidad. Rather than develop a haven from mosquitoes, they choose to create a double-skin wall that provides space for the insects. [Figs. 10.7 + 10.8] As the house's unusual form twists and bends, the mosquitoes within the double wall are brought closer and more deeply into the spaces of the house. François Roche wrote of the project's operation:

> Scenario: 1) Detection of the mosquito-borne West Nile Fever virus on the island. 2) Mixing this objective paranoia with a desire for safety. 3) Developing a Klein-bottle twist between the two contradictory data: humans and insects. 4) Living and dying of mosquitoes in the house trap. 5) Introducing a fragile structure and materials, like fabric netting

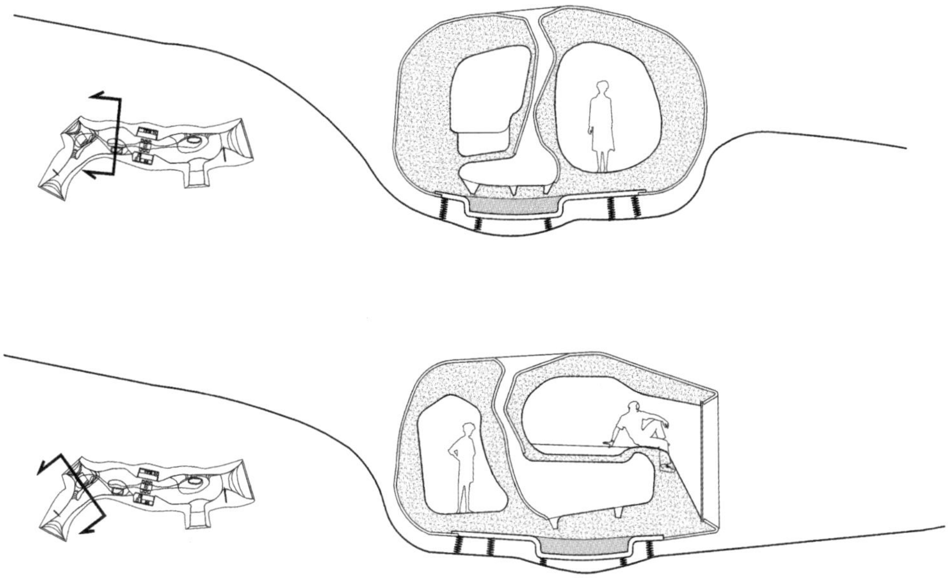

Fig. 10.7
Section diagrams of Mosquito Bottleneck, by R&Sie(n) Architects, Trinidad, 2003

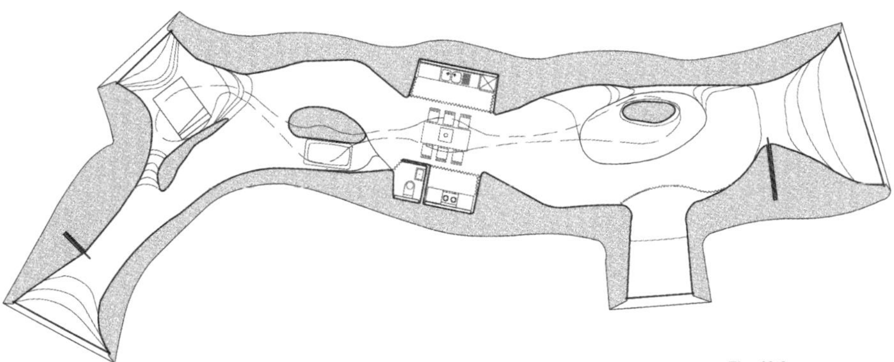

Fig. 10.8
Plan of Mosquito Bottleneck, by R&Sie(n) Architects, Trinidad, 2003

Fig. 10.9
Exterior view of Mosquito Bottleneck, by R&Sie(n) Architects, Trinidad, 2003

> everywhere, in recognition of the geographic position of this island, naturally protected against hurricanes. 6) Weaving together all the surfaces of the house—floor, facade and roof—with plastic wire and plastic shrink-wrap. 7) Resonance between the buzzing of the mosquitoes and the vibration of the structure.[8]

The project is not about mimicking the insect's forms of organization; rather, it embraces the metaphor of the insect as a new type of experience. The house suggests that domesticity is built on excluding fearful aspects of nature, while embracing others. But it also suggests new types of interpretations of one's personal natural milieu seen through an insectlike lens and projected onto the world. If insects were once cast out of the interior (or burned in effigy), here, in an admittedly rarified setting, they are brought closer to enable another concept of what an interior might be. This, in turn, suggests that the combination of insects, the interiors that house them, and the observation and inhabitation of such a space by people force a consideration of the very meaning of *human* and its interrelationship to these constructs. [**Figs. 10.9–10.13**]

Fig. 10.10
Skin detail of Mosquito Bottleneck, by R&Sie(n) Architects, Trinidad, 2003

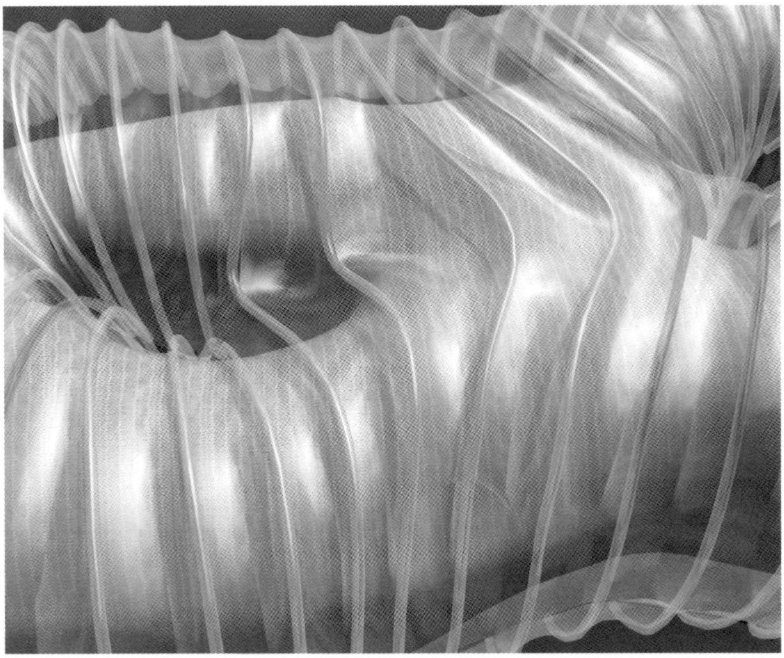

Fig. 10.11
(above, right to left)
Interior views of Mosquito
Bottleneck, by R&Sie(n)
Architects, Trinidad, 2003

Fig. 10.12
Interior of Mosquito
Bottleneck, by R&Sie(n)
Architects, Trinidad, 2003

Fig. 10.13
Interior detail of Mosquito Bottleneck, by R&Sie(n) Architects, Trinidad, 2003

Pigeons

Several subnatures in this book oscillate between desirable and undesirable. Within modernity, we often see key shifts where something as innocuous as smoke from a domestic fire eventually becomes a nuisance. The same is true for dankness, which was once literally constructed in garden grottoes, but many modernist tracts were devoted to eliminating it as a spatial quality. Once deemed undesirable, few subnatures shift back again to being wholly desirable. We don't have a "good" dust anymore, nor do we consider "good" exhausts or mud (at least of the spatial kind). Bugs, even the most beneficial ones, are unwelcome in most spaces. Such historical interpretive structures continue with architectural considerations of that most notorious of birds—the pigeon. Of all birds, pigeons retain a type of lesser status through their interactions with architecture in that they stain and infest buildings and transform the sounds emanating from them. We might even argue that the word *pigeon* takes on some of the power of the term *weed*, which people use broadly to describe unwanted vegetation. Many pigeons could be classified as doves or partridges; in fact, the distinction between pigeon and dove is quite arbitrary and might ultimately fall on the way a particular bird inhabits space. *Pigeon* is not so much a common name as a label of derision used to describe a member of the family Columbidae and the genus *Columba* that transgresses the imagined boundaries between human and animal worlds. But like many things under analysis in this book, this shift is very recent. Pigeons were once venerated in architecture, and of all birds, they contain one of the longest and most unique conceptualizations within architecture, invoking notions of sacrifice, class, liberty, and pestilence.[1]

If one looks past the antipigeon spikes on the entablatures of classical buildings, you might notice that virtually every classical edifice contains some reference to its occupation by pigeons. On the moldings of ancient and modern classical buildings, one often notices a pattern containing alternating ovoid forms and three small carved lines. This pattern, known as the "egg and dart," symbolizes the feet and eggs of pigeons. [**Fig. 11.1**] Temples, like modern structures, were homes to the flocks of pigeons that roosted in their nooks. Roosting pigeons were valued, and eggs laid in temples were sold as souvenirs and used for sacrifices.[2] Pigeons were also encouraged to breed inside buildings, and this extends from ancient to early-modern societies. The *pigeonnier* or dovecote, a building in which pigeons were cultivated for their eggs and meat, became a discernable architectural type in the West during the Middle Ages. Pigeonniers were

Fig. 11.1
Egg and dart motif on an architrave, 1777

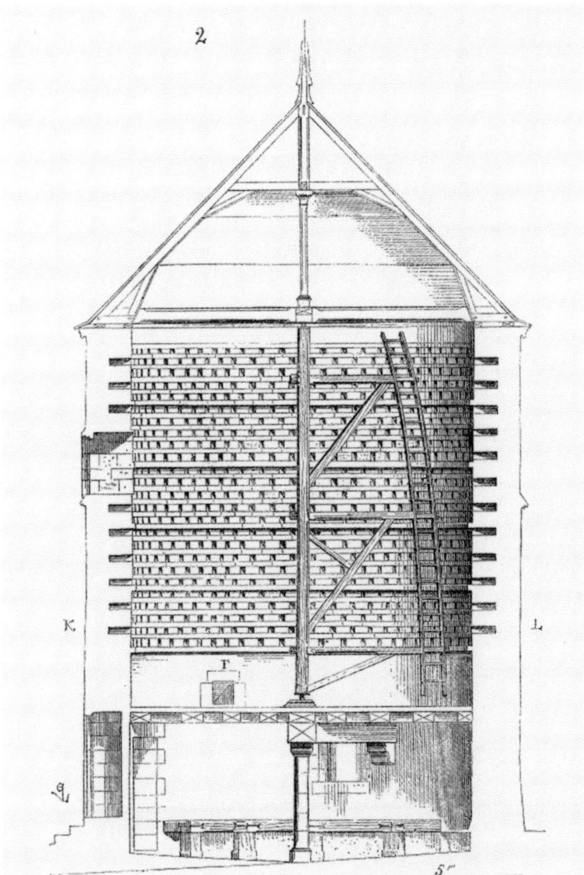

Fig. 11.2
Drawing of a *colombier* or *pigeonnier* from the *Dictionnaire Raisonné de l'Architecture Française*, Eugène Emmanuel Viollet-le-Duc, 1856

extensive utilitarian buildings typically built on large estates, and in the late eighteenth and early nineteenth centuries, they became fanciful buildings where images of architecture, gardens, and wild nature came together. Descriptions of the pigeonnier actually appear in works of architectural theory. For example, Leon Battista Alberti's treatise contains an entire section instructing readers on the methods for constructing buildings for pigeons.[3] Jean-Jacques Lequeu drew a section of a fanciful pigeonnier for a country estate. Within the *Dictionnaire Raisonné de l'Architecture Française*, Eugène Emmanuel Viollet-le-Duc noted the construction of the pigeonnier in medieval architecture, typically a round structure with a rotating ladder-assembly at its center. [**Fig. 11.2**]

Such a focus on pigeonniers within architectural theory might come as a surprise, but in Europe and its colonies, the construction of pigeonniers was governed by the state, making them a significant building type. In pre-revolutionary France, for example, the aristocracy retained exclusive rights to build pigeonniers. Viollet-le-Duc wrote one of the more evocative descriptions of the power relations moving through the Medieval pigeonner:

> Pigeonnier (dovecote): A building intended to house flocks of pigeons and permit them to brood, protected from the elements. During the Middle Ages the construction of a dovecote was a privilege reserved for feudal lords. The peasant could not have his own oven... neither was he permitted to have a pigeonnier. Pigeons, like herds of cattle and sheep, belonged to the lord, who alone could profit from them... The construction of the pigeonnier was thus an important affair. All castles possessed one or more pigeonniers; the manors, dwellings of the knights, small castles.... could also have a pigeonnier.[4]

According to Viollet-le-Duc, the number of pigeons that could be kept in a pigeonnier was linked to the amount of land owned by a particular member of the aristocracy. Additionally, the aristocratic classes forced nearby peasants to supply food for the pigeons raised in these buildings.[5] For these reasons, pigeonniers and pigeon trapping emerged as symbols of aristocratic privilege within early modernity.

In light of the above history, within nonarchitectural critiques of state power, the pigeonnier became an emblematic building. Charles Dickens noted the burdens on peasants in supplying food for pigeonniers in *A Tale of Two Cities*; Karl Marx commented on this and many other forms of alienation from nature within feudalism. To illustrate their critique of the governance over nature by capitalist states,

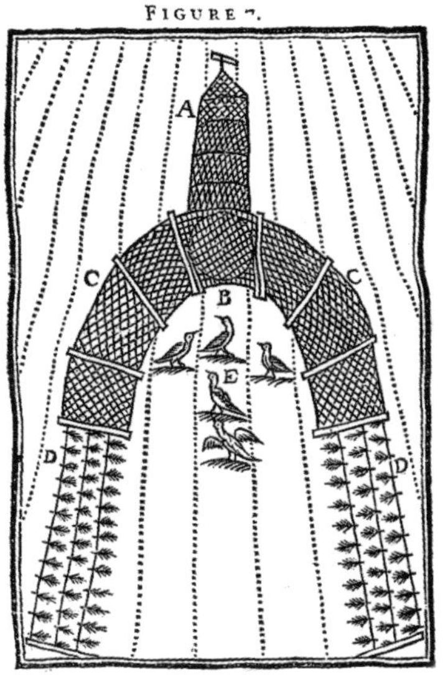

Fig. 11.3
Drawing of a bird trap by Noel Chomel, 1700s

Gilles Deleuze and Félix Guattari used an image of a crude bird trap designed by the eighteenth-century French writer Noel Chomel. Such traps illustrated a larger "apparatus of capture" that they believed took the form of stockpiling nature to impose economic control over the productivity of the earth and convert open territories into saleable land and human energy into labor for wage-bound forms of work.[6] [**Fig. 11.3**] As bird catching was intimately related to land ownership, the illustration by Chomel illuminates the broader concept of socionatural capture, but ultimately, the pigeon trap symbolizes the apparatus of capture, because we tend to associate birds, in particular, with a type of boundless negotiation of space by free nature.

The dialectic between a captured, state-managed nature and its potentially free form might enable us to see why, beginning in the mid-nineteenth century, images of unbound pigeons came to symbolize a progressive modernity. If pigeons were once the focus of various modes of state-governed capture for their meat, eggs, or navigation skills, then within emerging metropolitan culture, pigeons became images of a de-operationalized nature. Pigeons were admired for their ability to integrate into their surroundings and for their visual effects when flying. Images of pigeons hurtling through cities or populating squares became prevalent in modern writing, painting, and photography, particularly in the cities of Paris, New York, and especially Venice.[7] John Ruskin, writing in *The Stones of Venice*, went as far as to compare the type of buildings a society builds with the type of birds that populate it. Speaking of the pigeons (he called them doves) in the porch of St. Mark's, he contrasted the lively pigeon

Fig. 11.4
The procuracoes of Venice, from *The City of Tomorrow and Its Planning*, Le Corbusier, 1929

Fig. 11.5
O Pombal, by Oscar Niemeyer, Brasília, Brazil, 1960

with the more depressing birds that inhabited English buildings. Ruskin wrote:

> Between that grim cathedral of England and this, what an interval! There is a type of it in the birds that haunt them; for, instead of the restless crowd, hoarse-voiced and sable-winged, drifting on the bleak upper air, the St. Mark's porches are full of doves that nestle among the marble foliage, and mingle the soft iridescence of their living plumes, changing at every motion, of the tints, hardly less lovely, that has stood unchanged for 700 years.[8]

For him, the pigeons of Venice were metaphors for the relationship between architecture and social conditions. They are not a restless crowd but a well-integrated flock. Similarly, sixty years later, commenting on the scene at St. Mark's Square, Le Corbusier admired the pigeons as a type of natural module that worked well with the architectural scenography of existing buildings. [**Fig. 11.4**] He wrote fondly of a picture he took of pigeons swarming in the square:

> The uniformity of the innumerable windows in this vast wall on the Piazza San Marco gives the same play as would the smooth side of a room. The repetition of the same unit lends the wall a grandeur that is boundless but can be easily appreciated; the result is a type-form of a clear and simple nature. The pigeons of St. Mark's themselves add their own uniform module, providing a varied and effective note in the scheme.[9]

For Le Corbusier, nature occupied an ambiguous role as product of and counterpoint to the surrounding architecture and urbanism. But more so, nature and building shared a similar code, an idea that will be expanded on in contemporary forms of practice.

By understanding the historical role of pigeons as a form of free or captured nature governed by states, we might better understand one of the more enigmatic and modern structures specifically built for pigeons—O Pombal, in the Brazilian capital of Brasília (1960). Its architect, Oscar Niemeyer, produced a monument that evokes an obelisk in form but contains several ovoid openings that lead to a cratelike interior for pigeons. [**Fig. 11.5**] The monument's name, O Pombal, is a play on words that refers to the Portuguese term for pigeons, *os pombos*. As well as being a place for pigeons, it is also a reference to Portugal's interior minister and reformer Marquês de Pombal, who transformed and expanded both the territory and land privileges in colonial Brazil. In this concrete dovecote, Niemeyer provides a place for pigeons within the city of Brasília, but the pigeons are encouraged to roost, not for any use but for their symbolic power. Contemporary historian Justin Read argues that this structure is a symbol of postcolonialism; the pigeons may represent the people, and the dovecote becomes a Leviathan-like object that gathers a crowd, free of labor, as the pigeons are not harvested for their eggs or meat.[10] But more directly, we might see the collecting of pigeons (to no end) on the grounds of a state capital in a former colony as simply an inversion of colonialist law prohibiting the stockpiling of nature outside of the state. Niemeyer's dovecote physically enacts the privileges of a free nation to do as it will with its natural resources. Pigeons are gathered merely for the enjoyment of their flight, as notions of statehood and more cosmopolitan avian images conjoin.

All of the above architectural work with pigeons might surprise the contemporary observer. Today, we associate pigeons more often with forms of urban pestilence than with symbols of statehood. *Pigeon-infested* is a term we use to describe areas in cities and buildings that appear overrun with birds. Pigeons roost and relieve themselves on buildings, and the sounds they emit are often considered nuisances. In New York City, they are particularly reviled, labeled as *rats with wings* or *cockroaches of the sky*. They are known to damage forms of stone on urban buildings, and they are associated with respiratory diseases histoplasmosis and cryptococcosis. But despite the disgust they generate (and perhaps, in part, because of it) pigeons continue to be a relevant subject within recent architectural work. One of the most exemplary projects to engage pigeons on both material and historical terms is the Brooklyn Pigeon Project. The architectural firm Aranda\Lasch returns us to the dominant image of the pigeon as a form of free nature within modernity. They were influenced by concepts of flocking behavior from the mathematician Craig Reynolds and the omnipresence of these birds in New York City. The Brooklyn Pigeon Project attempts to both document the algorithms by which pigeons flock in cities and attempts to convey a vision of the city from the bird's perspective. The project embraces the image of the pigeon in modernity but uses it to produce a new perspective on urban space.[11] [**Fig. 11.6–11.8**] At virtually the same time as the Brooklyn Pigeon Project, the designer Matali Crasset revisited the concept of the pigeonnier in a project for the rural countryside of northern France. She developed a building to engage with local traditions of pigeon breeding—a tradition, as we have seen, that has existed for several hundred years. But Crasset makes the bright yellow pigeonnier a more visible object in an effort to retain the contemporary relevance of pigeon breeding. She enables the pigeons to become a functioning aspect of the local agricultural economy. [**Figs. 11.9–11.11**] The pigeonnier is designed so that farmers may collect the *guano* of the birds for fertilizer.[12] In effect, Crasset revisits the pigeonnier as the site of new strategies of capture and as an instrument of management and a local form of coordination. This conceit also extends into the recent work of Fritz Haeg, an architect and artist who develops "animal estates" for the sake of returning wildlife to cities. In several projects, Haeg proposes the introduction of hawks and other predators to manage urban pests such as pigeons. [**Figs. 11.12 + 11.13**] In many ways, images of pigeons in contemporary work remain caught between these tropes, valorized either as an unleashed nature or an object of spatial management. Following these recent projects, where does one go with the pigeon in architecture? This is obviously not a critical question, but as many of the chapters in this book illustrate, our contemporary interactions with nature seem stuck between a Victorian clean-up effort and a vitalist search for ways to transcend existing socionatural representations. Perhaps we might return briefly to the work of Niemeyer and his intimate understanding of statehood and territory in the appearance of this particular form of nature. How might we return this bird to architecture, retaining its subnatural history of territorial strife? If the pigeon is truly a relative label, akin to our discussion of weeds, we might ask not how architecture can incorporate pigeons, but how can architecture produce them as a form of urban life?

Figs. 11.6–11.8
(above and opposite)
The Brooklyn Pigeon Project,
by Aranda\Lasch, 2006

3 seconds of flight

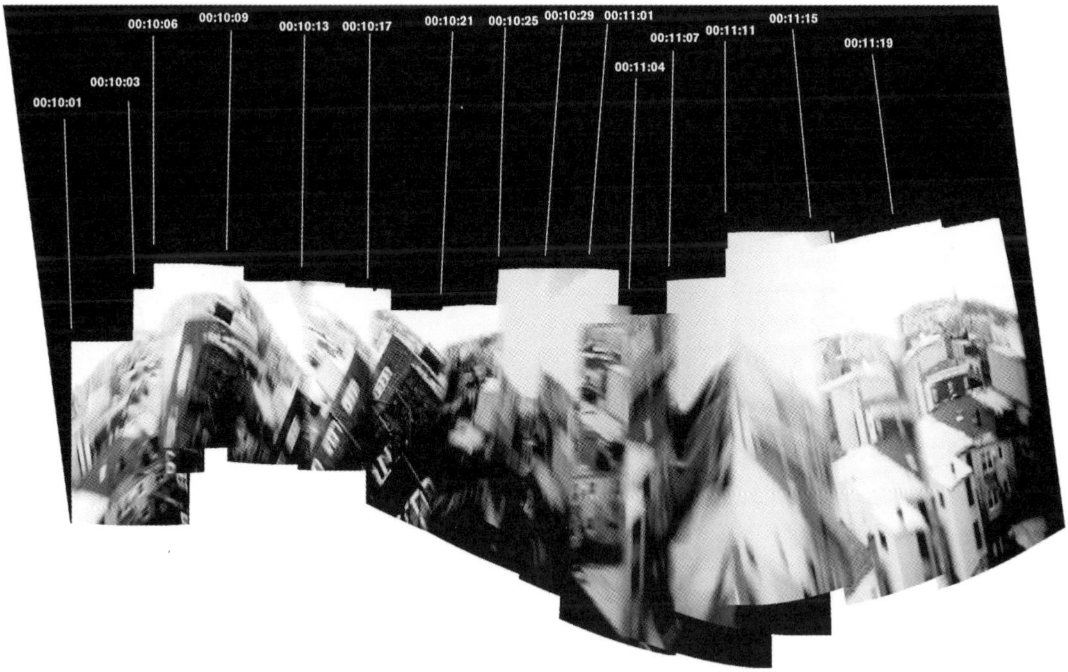

Figs. 11.9–11.11 (right and below) The Pigeon Capsule, by Matali Crasset, Cambrai, France, 2003

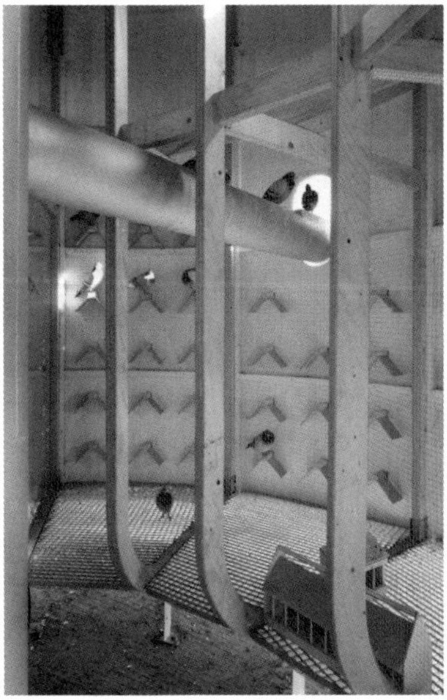

Figs. 11.12 + 11.13
(above and right)
Animal Estates, by Fritz Haeg, Whitney Museum of American Art, 2008

Crowds

We have explored atmospheres of gas and smoke, matters of dust and debris, and in our analyses of life we have investigated insects, weeds, and pigeons. We conclude by taking a closer look at the subnatures we produce as active participants in modern societies. Through our drive to urbanize, collect, and exchange ideas, we entangle ourselves within large gatherings in the world of the crowd, where subnatural forms also appear. Authors Victor Hugo and Elias Canetti have written of crowds as forms of nature akin to fire, water, or flows of sand.[1] In *The Hunchback of Notre Dame*, Hugo compared crowds assembled to seas, lakes, and rivers. He wrote of the crowds coursing through the city:

> The palace yard crowded with people looked like a sea, into which five or six streets, like the mouths of so many rivers, disgorged their living streams. The waves of this sea, increasingly swelled by new arrivals, broke against the corners of the houses.[2]

Elias Canetti wrote that the crowd could be described as series of nature metaphors:

> The manner in which fire spreads and gradually works its way round a person until he is entirely surrounded by it is very similar to the crowd threatening him on all sides. The incalculable movements within it, the thrusting forth of an arm, a fist or a leg, are like the flames of a fire which may suddenly spring up on any side.[3]

Within the work of these writers and others, the crowd appears as a form of flowing violent nature and as an engine of material, psychological, and political transformation. The transformative effects of crowds move through the history of architecture, where they take on new material properties. In works stretching from the nineteenth century to the present, the crowd appears as an agent of powerful social change, as in the works of Hugo and Canetti, but crowds also appear as more direct forms of subnatural matter. Without reducing human socialization, we may argue that crowds produced a type of socioclimate that fills spaces with new phenomena, such as odors, heat, gaseous substances, and intense noises. We might even consider whether these attributes manifest as significant aspects of the crowd's overall emotional and social effect. Crowds are formed with our bodies, which are significant generators of heat. Our body's internal temperature is almost 100 degrees

Fahrenheit, and the heat given off by us is equivalent to the heat of a 100-watt light bulb. In addition to heat, our bodies release moisture and carbon dioxide into the air through respiration and perspiration. The impact of our biological processes are most noticeable when our bodies are crowded together in theaters or assembly halls, and we can find depictions and writing documenting such struggles with the climate of the modern crowd stretching back over 150 years. Simply put, crowds are like nature, but they also launch new forms of nature akin to weather systems. They are climates made of people, and it is these transformative socionatural qualities that make crowds particularly alluring agents of social change and architectural conceptualization.

Institutional buildings provided the first venue in which architects and engineers sought to mitigate the interior nature of the crowd. Modern institutional structures such as hospitals, parliamentary halls, theaters, classrooms, and commercial exchanges all contained unprecedented assemblies of large crowds in relatively small spaces. In such spaces, emerging notions of bourgeois individuality rubbed against the clearly crowded spaces in which urban-middle and upper-middle-class life was staged. New types of experts, such as ventilation engineers, attempted to resolve the dichotomies between the crowded spaces of modern structures and a social milieu emphasizing the distinction of individuals.[4] The emanations from crowds involved both the actual temperature radiating from bodies and the gases emitted by people breathing. By releasing the heat of the crowd, ventilation engineers believed they could achieve varying levels of individual comfort, sanitation, and productivity. One of these early ventilation engineers, David Boswell Reid, explored how to expel the heat of the crowd in his design for the ventilation system at the Houses of Parliament in London. Reid developed a complementary ventilation structure that rivaled the actual meeting hall in scale. Composed of an enormous ventilation stack that used heat, filters, and moisture, the structure (Reid hoped) would evacuate expelled air from the hall. His system promoted the parliamentary hall as a space in which the atmospheric byproducts of human assembly were transformed, enabling government officials to withstand longer meeting times, but it was also symbolic of larger transformations of the inhospitable atmosphere of industrial urbanization.[5] The study of the assemblage of bodies in these types of spaces continued in the work of S. H. Woodbridge. In his work on the ventilation of schools and church buildings, Woodbridge developed some of the first building sections in which bodies were drawn as emitters of warmth and atmospheric gases. [**Fig. 12.1**] His drawings

Fig. 12.1
Drawing of crowds in a hall by S. H. Woodbridge, 1900

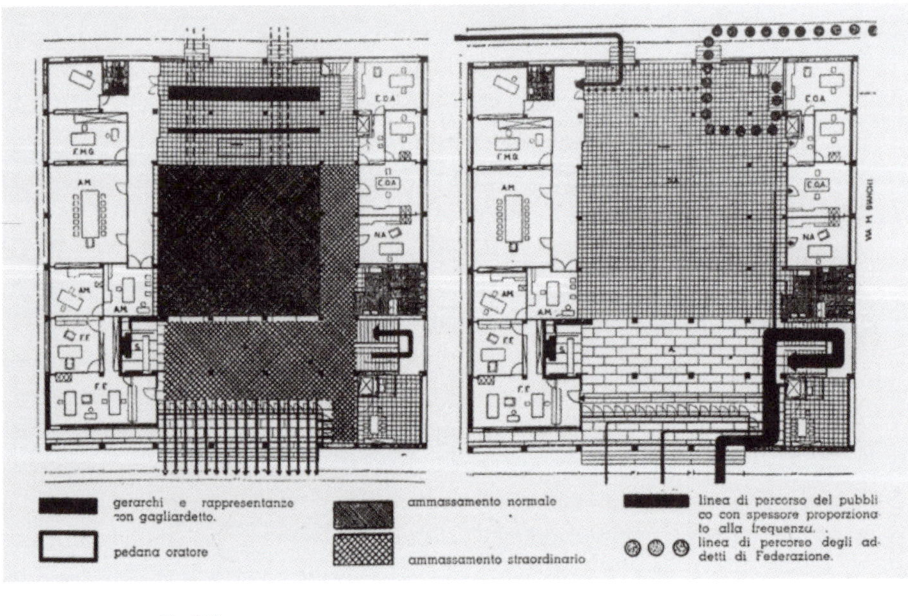

Fig. 12.2
Circulation diagrams, Casa del Fascio, by Giuseppe Terragni, Como, Italy, 1936

relate to the fears of lost individuality described earlier, and they touched on the emerging paranoia concerning the harmfulness of previously breathed—*vitiated* or *miasmatic*—air, terms influenced by the urban sanitarian reform movement, which attempted to mitigate the transmission of disease within building interiors.[6] By the late nineteenth century, American and European architectural engineers sought to reduce the effects of crowd heat in workplaces for the sake of productivity, comfort, and to dispel sanitarian fears. Building engineers studied a range of workplaces, from the indoor climates of factories to trading halls, to determine how to design environments that eliminate the impact of bodily fatigue.[7]

If nineteenth-century architectural engineers were concerned with the atmospheric emissions of crowds, modernist architects were equally concerned with the circulation of people, the ability to distribute information through crowds, and the sheer masses of people that might be assembled within a single enclosure. As the historian Adrian Forty has pointed out, the concept of *circulation* provided a scientific metaphor in modern architecture that linked the natural movement of fluids to the movement of people in space. The circulatory metaphor influenced discussions of crowds, both in terms of the movements of masses of people and their stagnation or congestion. Crowds might be understood as the result of poor spatial planning, as blockages in flows. In addition to circulatory concepts, crowds were conceptualized relative to the transparency they afforded and the possible masses that modern engineering might accommodate in a single space. A series of images stretching from the early 1930s to the late 1960s—addressed to significantly different political ends—illustrate some key crowd concepts in the history of the modern movement in architecture. In his design for the Fascist headquarters, the Casa del Fascio, the Italian Rationalist Giuseppe Terragni developed a diagram depicting the varying intensities of the crowds that he envisioned inside the building for the political rallies of Mussolini. [**Fig. 12.2**] Terragni's crowd diagrams associate the extremes of crowds with a type of opacity; he wanted the leaders of the state to be seen and heard by the assembled masses. His diagram contrasts the circulatory areas of stairs, drawn with vector diagrams, with the main assembly area on the ground floor, illustrated with hatching to represent varying levels of crowd massing.[8] A slightly different concept of the crowd appears in the work of Ludwig Mies van der Rohe. For Mies, the crowd's power was achieved through structural and spatial ingenuity—a concept best visualized in his design for an assembly of 50,000 people at the Chicago Convention Center. The collage presentation for the Convention Hall includes Mies's image clippings from

Fig. 12.3
The Convention Hall,
by Ludwig Mies van der Rohe,
Chicago, Illinois, 1953

Fig. 12.4
Cloud Nines, R. Buckminster
Fuller, 1962

Life magazine's coverage of the 1952 Republican Convention. [Fig. 12.3] The collage's peculiar mixture of nationalistic exuberance and the roof's vast steel-structural assembly suggests a crowd brought into being by America's postwar political and technological ascendancy.[9] By the 1960s, the various modernist concepts of the crowd—its engineered volume, atmospheric features, circulatory elements, and collectivity—became conjoined in one of the more alluring images produced by architect and engineer R. Buckminster Fuller. In Cloud Nines, his project for a floating city, Fuller hypothesized that if the interior air of an enormous one-mile-diameter geodesic sphere, housing an entire metropolis, was warmed one degree, the structure would begin to float. A floating aggregation of people would literally be able to move around the earth, as one giant collective. In an earlier description of the idea, Fuller spoke of the dome's air being heated by the sun; yet, in a 1962 version of the project, he illustrated denser, more opaque surfaces for it. [Fig. 12.4] Were these domes to be warmed by the masses of people within them? If crowds in the nineteenth century were conceptualized as generators of atmospheres, and if crowds in the mid-twentieth century were conceptualized as a type of engineered collective, here the collective literally becomes part of the earth's atmospheric system. Crowd, structure, and atmosphere coalesce.

By the late 1970s, new concepts of the crowd were being formulated in architectural and engineering theory in which modernist themes and concerns were revisited. The Dutch architect Rem Koolhaas arrived at some of the most exciting conceptualizations and images of crowds as catalysts for entirely new forms of urban architecture. Beginning with his 1978 *Delirious New York* and extending into later work such as *S, M, L, XL*, Koolhaas interrogated the possible outcomes of crowds as generators of new forms of programmatic experience in architecture. In his earlier work, which set the tone for more recent experiments, Koolhaas was reacting against the progressivism that urbanists and architects were directing against the metropolitan conditions of crowds and density. At a time in which urbanists were struggling to dissipate the density of urban life—in the name of humanity—Koolhaas was striving to enhance and exaggerate the crowd's potential.[10] In what might be the engineering analog for Koolhaas's culture of congestion, in the late 1970s and into the 1980s, progressive engineering firms were increasingly considering the climatic effects of congestion on building interiors. In what they termed *internal loading*, building engineers examined what happens when a building's climate tips past a balance point, when the internal loads generated by human bodies and computer equipment completely overwhelm the external load of a building. In other words, *internal loading* refers to instances in which the weather of a building is generated within it, versus outside of it. When we think of the climatic nature acting upon spaces, we tend to consider the intensity of the sun, the speed of the wind, and air temperature. All of these factors impact the climatic parameters of architecture, but in addition to the external climate, as we have seen, crowds within a building can generate an equally robust and problematic internal climate, and in some cases the internal climate dominates the impact of the sun, wind, and external air temperature. While some firms simply sought to air-condition internally loaded buildings, in the 1980s the engineering firm Arup explored methods in which internal loading by aggregations of human bodies are incorporated directly into a building's climate systems as agents of less energy-intensive climate engineering schemes. Engineers at Arup conceptualized workers' bodies and computer heat, feeding the energy recovery systems into the interior of a modern office building.[11]

The conceptual link suggested here—between late-modern architectural theories of aggregation, density, and crowding and an architectural science that explores the dynamics of bodies gathered in interior space—unite in a project by the firm Décosterd & Rahm. In the Omnisports Hall, a recreational facility in Neuchâtel, Switzerland, the firm worked with scientists to study the gaseous emissions of the human body and consider how this human-produced nature might be circulated through buildings to generate a new type of architecture. [**Fig. 12.5**] The building contains a central indoor sports area for basketball, running, and soccer surrounded by spaces that absorb the warmth and gas emanating from the activities of the players. [**Figs. 12.6 + 12.7**] Within the surrounding space, the architects propose the cultivation of plants capable of replenishing the nutrients lost by players; the entire structure is a type of ecosystem of crowded bodies. The architects write:

> This architecture is a chemical and biological reformulation of environmental space, accomplished via transpiration and photosynthesis, combustion and respiration, in which man assumes his place physiologically... heat and oxygen are absorbed by the players, who in turn provide the carbon dioxide and water vapor lost by the body as it transforms the chemical energy of the simple substances assimilated during digestion into kinetic energy. The resultant vitiated air migrates to the side windows, where it condenses. The plants between the panes absorb carbon dioxide from the air, the condensation produced by

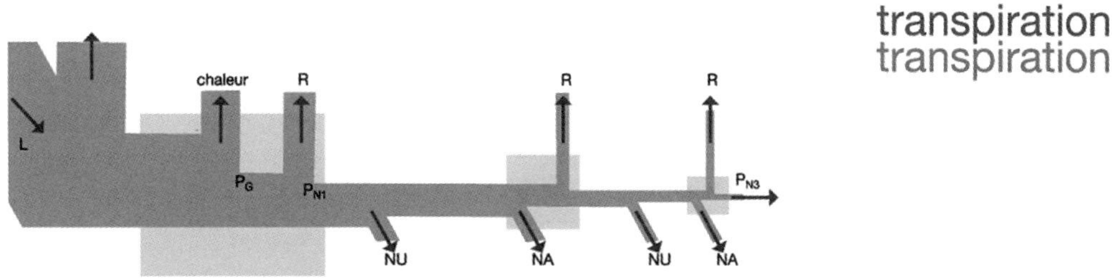

	L	L_A	P_{N1}	P_{N2}	P_{N3}
kcal	3000	1500	15	1,5	0,15
kJ	12600	6300	63	6,3	0,63

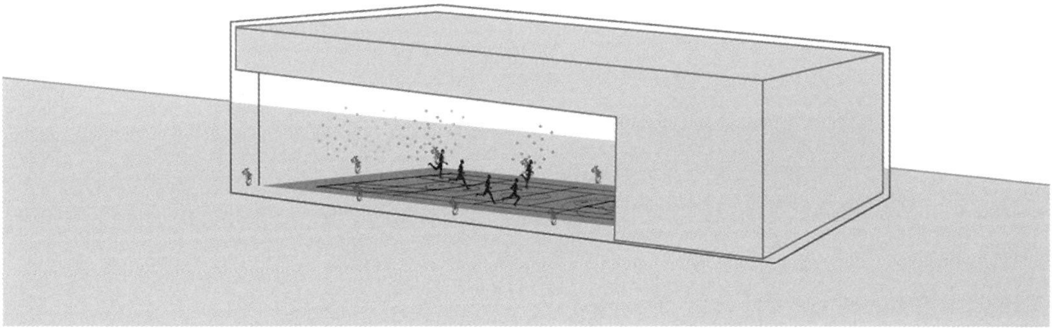

Fig. 12.5
Competition entry, Omniports Hall, by Décosterd & Rahm, Neuchâtel, Switzerland, 1998

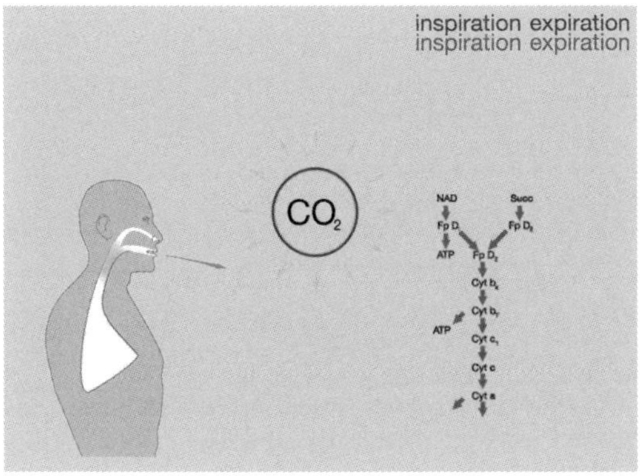

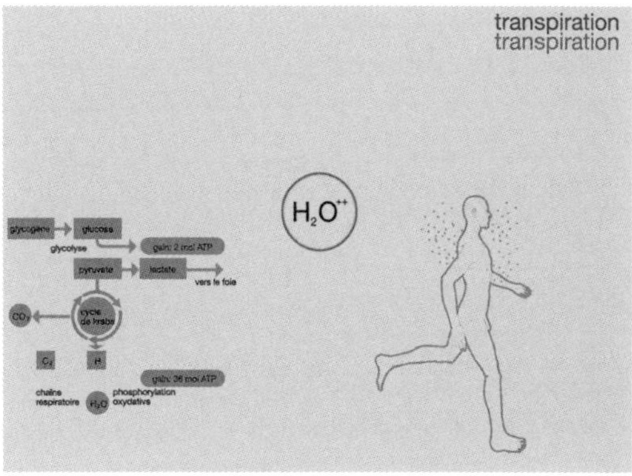

Figs. 12.6 + 12.7
Carbon dioxide and hydrogen dioxide diagrams, Omnisports Hall, by Décosterd & Rahm, Neuchâtel, Switzerland, 1998

> transpiration from the players, and their mineral salts. This chlorophyll photosynthesis serves to transform solar energy into nutrients and to produce the oxygen required by the players to burn the foods they need in order to release energy.[12]

Through Décosterd & Rahm, we see how crowded buildings are increasingly viewed as structures having a thousand small suns and a thousand vaporous winds. But what can be made of this living nature has yet to be realized in this particular context. Décosterd & Rahm identify the crowd as an ecological instrument as opposed to a tool of social agitation. [**Figs. 12.8 + 12.9**]

In considering the particular "nature" of crowds, their historical role as spaces of social change, and architects' future manipulation of these facets, we should consider one final project—Open Columns by the architect Omar Khan. In this project, Khan developed a series of flexible conical chambers that drop and expand when carbon dioxide levels increase as a result of the crowding of inhabitants in space. The project functions as a responsive form of architecture, with sensors in tune with the breathing cycles of a building's inhabitants. [**Figs. 12.10–12.14**] As the columns drop, they inhibit the assembly of crowds, thus functioning as a form of crowd control, albeit in the name of maintaining human health. Like Décosterd & Rahm, Khan views crowds as a form of nature, but here the crowd's possible dangers are architecturally visualized and ultimately mitigated. Both Décosterd & Rahm and Khan respond to the chemistry of the crowd, but the possible relationship between the nature of the crowd and its historical role remains to be staged. In recognizing the subnatural character of the crowd, their work forces us to consider how architects might employ the crowd's socionatural chemistry in an era in which the transformative aspects of crowds appear momentarily suspended.

Figs. 12.8 + 12.9 (opposite, top and bottom) Perspective renderings, Omnisports Hall, by Décosterd & Rahm, Neuchâtel, Switzerland, 1998

Fig. 12.10
Carbon dioxide diagram, Open Columns, by Omar Khan, 2007

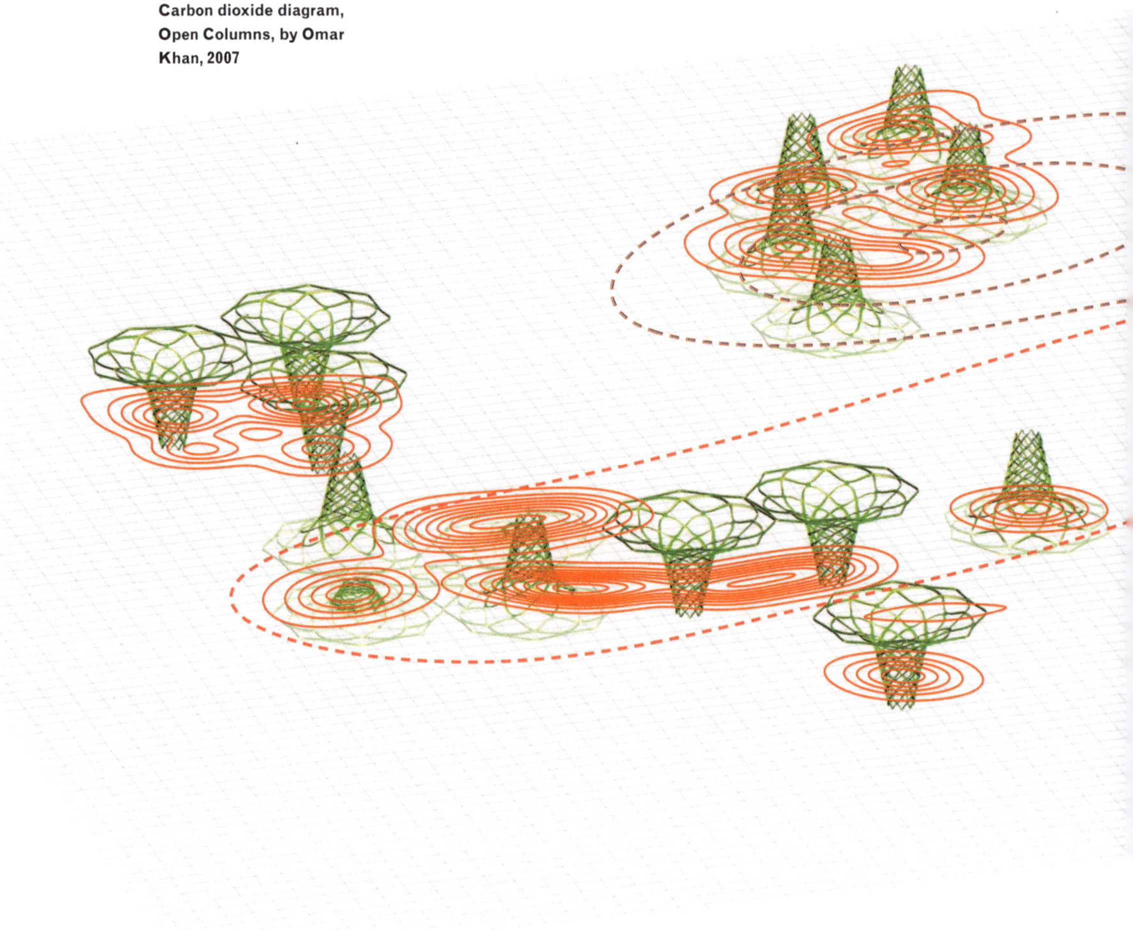

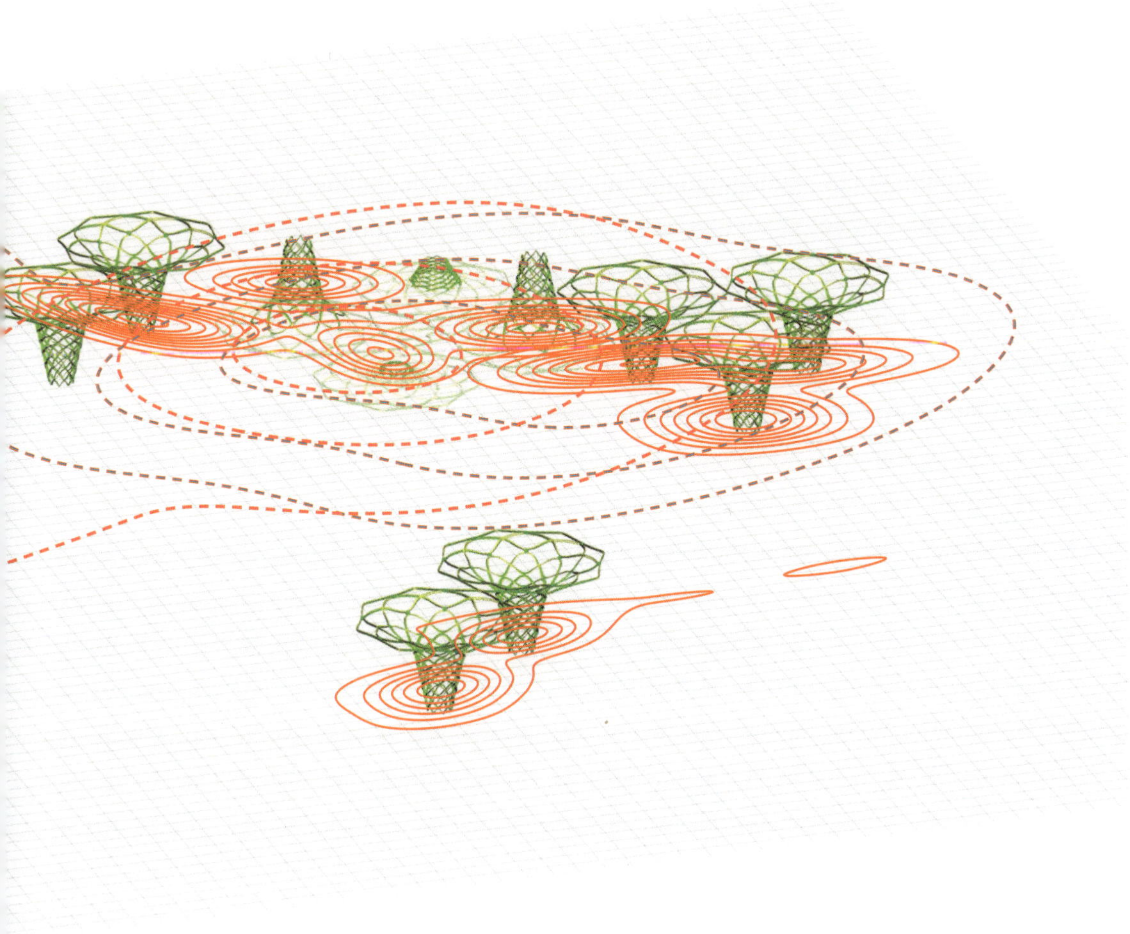

Fig. 12.11
Aggregations of columns, Open Columns, by Omar Khan, 2007

Figs. 12.12 + 12.13
(opposite, top and bottom) Columns deployed in response to increased carbon dioxide from respiration, Open Columns, by Omar Khan, 2007

Fig. 12.14
(above) Deployed columns in crowd, Open Columns, by Omar Khan, 2007

Weeds

1. See Mary Douglas, *Purity and Danger: An Analysis of Concepts of Pollution and Taboo* (London: Routledge, 1966); and for the architectural implications of her philosophy, see Ben Campkin and Paul Dobraszczyk, "Introduction," in "Architecture and Dirt," special issue, *Journal of Architecture* 12, no. 4 (2007): 347-51. On weeds, see the introduction in Clinton L. Evans, *The War on Weeds in the Prairie West: An Environmental History* (Calgary: University of Calgary Press, 2001).

2. Marcus Pollio Vitruvius, *The Ten Books on Architecture*, 104-6.

3. Richard Payne Knight, *The Landscape* (1794): lines 228-35.

4. On a general overview of the Picturesque, see Robin Middleton and David Watkin, *Neoclassical and 19th Century Architecture* (New York: Electa/Rizzoli, 1987), 37-46. For a more politicized interpretation, see Ann Bermingham, "System, Order and Abstraction: The Politics of English Landscape Drawing around 1795," in *Landscape and Power*, ed. W. J. T. Mitchell (1994; Chicago: University of Chicago Press, 2002), 77-102.

5. Adrian Forty, *Words and Buildings: A Vocabulary of Modern Architecture* (New York: Thames and Hudson, 2000), 155-57.

6. George Gessert, "Bastard Flowers," *Leonardo* 29, no. 4 (1996): 291-98.

7. Louis H. Sullivan, *Kindergarten Chats and Other Writings* (1902; New York, Dover, 1979), 86.

8. Ibid., 143.

9. On this weed imagery, see Lauren S. Weingarden, "Naturalized Nationalism: A Ruskinian Discourse on the Search for an American Style of Architecture," *Winterthur Portfolio* 24, no. 1 (1989): 43-68.

10. On the rhizome concept see the introduction to Gilles Deleuze and Félix Guattari, *A Thousand Plateaus: Capitalism and Schizophrenia, trans. Brian Massumi* (Minneapolis: University of Minnesota Press, 1987).

11. Cero9 architects, Magic Mountain project text, unpublished.

12. R&Sie(n) Architects, nMBA project text, unpublished.

13. From the architect's website.

Insects

1. Juan Antonio Ramiréz, *The Beehive Metaphor: From Gaudí to Le Corbusier*, trans. Alexander R. Tulloch (London: Reaktion, 2000).

2. Quoted in Greg Lynn, "Body Matters," in "The Body," special issue, *Journal of Philosophy and the Visual Arts*, ed. Andrew Benjamin (1997): 62.

3. Ibid., 62.

4. Ibid., 61-69.

5. On field theory, see Sanford Kwinter, "La Cittá Nuova: Modernity and Continuity," in *Architectural Theory*, ed. Harry Francis Mallgrave (Malden, MA: Blackwell, 2008): 474-75; and Stanley Allen, "From Object to Field," in *Architecture After Geometry*, ed. Maggie Toy (1999), 24-31. On the use of termite mounds and their influence on engineering practices, see David Gissen, *Big and Green*.

6. See Cristopher Hollingsworth, *Poetics of the Hive: Insect Metaphor in Literature* (Iowa City: University of Iowa Press, 2001).

7. See Gilles Deleuze and Félix Guattari, *Kafka: Toward a Minor Literature*, trans. Dana Polan (Minneapolis: University of Minnesota Press, 1986); *A Thousand Plateaus*, 232-309; Donna Haraway, "A Cyborg Manifesto: Science, Technology, and Socialist-Feminism in the Late-Twentieth Century," *Simians, Cyborgs, and Women: The Reinvention of Nature* (New York: Routledge, 1991), 149-81.

8. François Roche, project text, unpublished.

Pigeons

1. For an overview of pigeons in history see Andrew Blechman, *Pigeons: The Fascinating Saga of the World's Most Revered and Reviled Bird* (New York: Grove Press, 2006).

2. George Hersey, *The Lost Meaning of Classical Architecture: Speculations on Ornament from Vitruvius to Venturi* (Cambridge, MA: MIT Press, 1988), 34.

3. Leon Battista Alberti, *The Ten Books of Architecture* (New York: Dover, 1485/1755), 103.

4. See the entry "Colombier," Eugène Emmanuel Viollet-le-Duc, *Dictionnaire Raisonné*, trans. David Gissen and Molly Slota (Paris: B. Bance, 1854).

5. A surprisingly concise piece on these buildings and their politics can be found in Alice Furlaud, "Homing in on Pigeon Towers," *New York Times*, April 1, 1990.

6. Deleuze and Guattari, *A Thousand Plateaus*, 424-73.

7. In particular, see the work gathered under the banner of the Ashcan School in the early twentieth century United States.

8. John Ruskin, *The Stones of Venice, second volume* (London: Smith, Elder and Co., 1873), 67.

9. Le Corbusier, *The City of Tomorrow and Its Planning* (New York: Dover, 1987), 69.

10. Justin Read, "Alternative Functions: Oscar Niemeyer and the Poetics of Modernity," *Modernism/modernity* 12, no. 2 (2005): 253-72.

11. See "Flocking," in Benjamin Aranda, *Tooling* (New York: Princeton Architectural Press, 2005). 62-74.

12. Matali Crasset, *The Pigeon Loft* (Paris: Pyramyd, 2004).

Crowds

1. Elias Canetti, *Crowds and Power*, trans. Carol Stewart (1962; New York: Viking Press, 1986).

2. *Crowds*, ed. Jeffrey Schnapp and Matthew Ties, (Stanford, CA: Stanford University Press, 2006), xiii.

3. Canetti, *Crowds and Power*, 27.

4. Dell Upton, *Architecture in the United States* (Oxford: Oxford University Press, 1998), 141- 47.

5. See Reyner Banham, *The Architecture of the Well-Tempered Environment* (Chicago: University of Chicago Press, 1969); and Robert Bruegmann "Central Heating and Forced Ventilation: Origins and Effects on Architectural Design," *The Journal of the Society of Architectural Historians* 37, no. 3 (1978): 143-60.

6. S. H. Woodbridge, *Upwards Versus Downwards Ventilation* (London: Robert Boyle and Son, 1900).

7. See the section on ventilation and air-conditioning in Cecil D. Elliott, *Technics and Architecture: The Development of Materials and Systems for Buildings* (Cambridge, MA: MIT Press, 1992).

8. See Thomas L. Schumacher, *Surface and Symbol: Giuseppe Terragni and the Architecture of Italian Rationalism* (New York: Princeton Architectural Press, 1991), 145.

9. On the images moving through this particular collage, see Neil Levine, "The Significance of Facts: Mies' Collages Up Close and Personal," *Assemblage* 37 (1998): 70-101.

10. See Rem Koolhaas, "Life in the Metropolis or the Culture of Congestion," *Architecture Theory Since 1968*, ed. K. Michael Hays (Cambridge, MA: MIT Press, 1998), 320-31.

11. See the essay by Guy Battle in David Gissen, *Big and Green*.

12. Philippe Rahm, project text, unpublished.

Epilogue

Imagine an architect looking at the subnatures of the city: a cloud of smoke, a pool of mud, a pile of debris. He or she might see such things as emblematic of mismanagement, abandonment, or catastrophe. These forms of subnature often appear threatening to a city's social and physical fabric, and this architect might try to position his or her work to control or purge this matter, removing its presence and memory from the city. On the other hand, perhaps this architect is not troubled by subnatures, understanding them to be fundamentally useless and ignoring them in favor of those forms of nature that can be incorporated into a building's systems. Such an architect overlooks the problematic matter in the environment and instead focuses on those aspects of the environment that enhance buildings as technical systems. But, reflecting for a moment, perhaps this hypothetical architect considers these strange forms of nature as a material endemic to architecture and cities, as opposed to an aberration that must be consolidated, removed, or dismissed. He or she is not only engaged with the realities of the modern world but with the social processes that surround architecture, urbanism, and history. To rid cities of subnature negates aspects of urbanity while advancing a narrow concept of architecture's proper environment. By seeing only those things that are useful to a building's program, an architect dismisses key aspects of contemporary urban life. Furthermore, to simply ignore these forms, to demand autonomy from them or embrace them without any concern or understanding of their problems is altogether blind and irresponsible. Ultimately, the hypothetical architect might consider engaging these discomforting things in a state of reflection, acknowledging the genuine difficulties people face when confronted by the city's hostile matter. By actively reflecting on the alienating material of the socionatural environment, we might as a profession arrive at a truly radical and alternative concept of the environment for the contemporary architect.

When we talk of architecture engaging with the environment, very often we mean to say that architecture is harmonizing with, or open to, some aspect of an uncorrupted nature. An architecture that engages with the environment usually incorporates or mimics the mechanics of trees, sunlight, water, and wind; whether developing a country house or skyscraper, the architect attempts to work the form, program, and systems of the building into a mutually beneficial relationship with the environment. Nature and architecture support one another and leave each largely untransformed. But as this book has demonstrated, the environment is much more than the nature we often imagine to be in some prehuman and pristine form; it is composed of subnatures produced by social, political, and architectural processes and concepts. Unlike the natural environment, we cannot possibly imagine a subnatural environment generated by, nor found within, a nonhuman world. Subnatures force us to confront the implicit nonsocial character of nature, as it is invoked in discussions of architecture and the environment.

Not only is the subnatural environment made by society, the cruel irony is that, unlike other effects of civilized life, subnature cannot be used as a stable context or material for architecture. Its inherently uncontrollable, filthy, and fearsome aspects confront the stability of architecture itself. To bring subnature into architecture transforms our expectations of it. If we consider one of the buildings explored in this book—the B_Mu Tower by R&Sie(n) Architects—we see that the architect engages the building with the city's dust-ridden atmosphere, forcing us to question the concept of architecture as refuge from the urban environment. Its proposed program and structure exists in tension with the landscape's polluted context, even as the architect attempts to establish a rapprochement with it. As the exhaust becomes part of the building, it appears ever more alien and frightening. Such a building demonstrates the constraints that subnature forces on the architect. But by entering into this dialectic with subnature, the architect actively produces knowledge about the subnature of the city, our fears of it, and what we expect from an architecture that engages with its surroundings. At the very least, such a building enables the constituent features of nature to be understood, debated, and perhaps ultimately transformed, while leaving a record of an earlier struggle.

By forcing us to rethink the environment's nonhuman and benevolent qualities, subnatures urge us to rethink ideas of use tied to that environment. Within a capitalist economy, the nature outside of society appears as a resource, whether we speak of a pine tree that is transformed into building lumber or a giant sequoia tree that people travel thousands of miles to visit in a national park. In contrast, the inherent resistance of subnatures, such as mosquitoes, dust, or smoke, to notions of usability and mass consumption positions them as a different type of nature. Certainly, a subnature can be transformed into a spectacle or commodity, as with more naturalistic forms of nature, but in general terms, subnatures force us to confront our prevailing relationships to the environment. This is particularly the case when we consider them in relation to integrated urban systems and building infrastructures and technologies. Consider, for example, the types of nature that municipal officials, developers, and architects typically incorporate into urban spaces and buildings. Western municipal managers and their architects have a tendency to marginalize any aspect of the environment that cannot be operationalized for the smooth functioning of the city in favor of forms of nature in which some potential cash benefit can be assessed, whether directly or speculatively. In New York City, the meager funds for "public nature" are used to maintain water systems and those parks that support its neighboring real estate. While all cities require robust water and park systems for the health and well-being of their inhabitants, environments that stand in contrast to these systems are constantly under threat, such as parks operated in abandoned lots, pigeon lofts, chicken coops, and other minor forms of nature, often found in struggling, poorer neighborhoods. More recently, desires for a systematized and potentially profitable urban nature are carried into the realization of many contemporary green buildings and neighborhoods. Here, too, the idea of the natural environment as feeding an infrastructural system dominates, and those natures that cannot be incorporated into building systems or urban networks are simply negated or put under threat.

While subnatures can be transformed into forms of spectacle, they cannot easily be made into functional instruments, as in the networked fantasies of green cities or the air systems of green office buildings. It is highly unlikely that cities will ever have networks for the distribution of dust or smoke. Of course, some subnatures, like any form of matter, can be transformed into resources, but doing so forces us to question our own ideas about what resources offer us. When subnatures become integrated into contemporary building systems, as in the case of Philippe Rahm's Omnisports Hall or Underground Houses, they exist in tension with the normative subtexts of particular building systems. In the Omnisports Hall, the incorporation of unpleasant bodily exhalations (carbon dioxide and sweat) into the building system compells us to reconsider notions of individuality and physical propriety. In his Underground Houses, the dankness of the earth is used as a form of passive cooling, but in pulling this air out of the earth, Rahm opens up a strange new zone in the house: a large and mysterious underground chamber. He uses the dank air from the chamber to reorganize the typical programmatic relationships of a middle-class house. Unlike more normative forms of nature, subnatures are not a means for making existing buildings and cities into better functioning wholes; they are critical instruments for examining how our notions of the environment either support or undermine existing experiences of architecture and urbanism.

Subnatures enable us to better understand our environment as a product of social and historical processes, as something tied to social history, as much as natural history. But even as we begin to understand all of nature as bearing the imprint of human influence, the desire for a natural architecture (with all of its paradox) continues to appeal to us. The naturalistic and mechanistic concept of the environment critiqued here appears anew, fused in contemporary fantasies of a futuristic ecological architecture. Such buildings, by architects such as MVRDV, Terreform, or Guallart Architects, represent cutting-edge thinking on architecture-nature dynamics. These firms, and others, offer fantastical concepts and images of society generating new bonds between nature and architecture. They advance a futuristic natural environment as the site for an architecture integrated within or emerging from nature. In some projects, droughts, floods, animals, and vegetation overtake the world, and a corresponding architecture attempts to situate itself within this new reality. While original, this work advances existing and much less radical notions of the relationship between architecture and the environment, and though it proposes a thrilling aesthetic, the work precludes a more critical understanding of nature's role as a reductive, mechanically invoked resource.

In contrast, subnatures often operate within social and architectural frameworks, more so than technical or geological ones. Consider the Irish Pavilion by Tom dePaor Architects, in which a strange new form of architecture emerges directly from a historical understanding of mud. The power of this engagement is not tied to geology or physics but to the bog mud's role as an aspect of Irish national consciousness. Similarly, Manuel Herz and Eyal Weizman's pavilion designs for their proposed Open-Air Parkcafé in Cologne, Germany, invoke the forms of natural houses with their turfed-over forms. But the roofs are actually formed out of soil obtained from the debris fields of a post–World War II landscape. The earth used within this project is first and foremost a historical material. In some cases, such as in the work of Jorge Otero-Pailos, subnature becomes a form of historical inquiry in itself. The preservation of smoke and dust becomes a new kind of commentary on history. It is this historical aspect that makes subnature such a useful contribution to contemporary debates.

Dankness, mud, gas, insects, and the other subnatures discussed in this book may at times appear threatening to architecture and its inhabitants, but subnatures are not only powerful because they are frightening or abject; rather, the notion of subnature forces us to confront existing environmental concepts and architectural techniques.

These confrontations come with benefits and dangers that must also be assessed and considered before subnature can emerge as a substantial concept. As they are often associated with the breakdown and denigration of the enviro-social sphere, we must not use them to advance a simplistic, picturesque nihilism of the present. Subnature should not be used to give in to the processes of pollution and war that appear to generate many horrifying landscapes. Nor is subnature meant to promote urban decay or celebrate infrastructural breakdowns. It is not intended as a source of fantasy, in which architects provoke their peers by drawing thousands of smokers or feral animals into the spaces we inhabit. That is not the point of the preceding chapters. Whether examining foul forms of nature, disasters, or war, all of the architects featured in this book qualify their interest in landscapes produced by power and inequity. Ultimately, this book promotes a concept of the environment that lacks the passivity and asocial qualities attributed to architecture's natural environment. It challenges the reductive and naturalistic aspects of contemporary architecture-nature dynamics, promoting a dialectic that radically rethinks both architecture and nature, and it helps us understand how certain ideas about nature are historically driven, socially and architecturally. I cannot argue that subnature will make peoples' experiences of architecture more pleasant or peaceful, but I do believe that it provides us with the intellectual and historical conceptualization of nature that many architects and architectural theorists have craved. Ultimately, *Subnature* is not about what is natural to architecture; it is about the natures we produce through our most radical architectural concepts.

Alberti, Leon Battista, The *Ten Books of Architecture*. New York, Dover, 1485/1755.

Allen, Stanley, "From Object to Field" in *Architecture After Geometry, Architectural Design* (1999): 24-31.

Amato, Joseph, *Dust: A History of the Small and the Invisible*. Berkeley and Los Angeles, University of California Press, 2000.

Aranda, Benjamin, *Tooling*. New York, Princeton Architectural Press, 2006.

Bachelard, Gaston *The Poetics of Space*. New York, Beacon, 1994.

Banham, Reyner, *The New Brutalism*, 1966.

——. *The Architecture of the Well-Tempered Environment*. Chicago and London, The University of Chicago Press, 1969, 1984.

——. *Megastructure: Urban Futures of the Recent Past*. London : Thames and Hudson, 1976.

Bermingham, Ann "System, Order and Abstraction: The Politics of English Landscape Drawing around 1795." In W.J. T. Mitchell, *Landscape and Power*. Chicago, University of Chicago Press, 2002: 77-102.

Benevelo, Leonardo, *History of Modern Architecture*, Cambridge, Mass, Mit Press, 1977.

Blechman, Andrew, *Pigeons: The Fascinating Story of the World's Most Revered and Reviled Bird*. Grove Press, 2007.

Bruegmann, Robert "Central Heating and Forced Ventilation: Origins and Effects on Architectural Design", *The Journal of the Society of Architectural Historians*, 37:3 (1978): 143-160.

Byles, Jeff, *Rubble: Unearthing the History of Demolition*. New York, Harmony Books, 2005.

Cadava, Eduardo, "Leseblitz: On the threshold of violence" *Assemblage*, 20 April (1993): 22-23

Campkin, Ben and Dobraszczyk "Introduction" special issue on "Architecture and Dirt." *Journal of Architecture*. Volume 12:4 (2007): 347-351

Campkin, Ben and Rosie Cox *Dirt: New Geographies of Cleanliness and Contamination*. London, I.B. Taurus, 2008.

Canetti, Elias, *Crowds and Power*. New York, Noonday, 1984.

Carapetian, Michael Interview with the author. June 12, 2008

Cohen, William and Johnson, Ryan *Filth: Dirt, Disgust and Modern Life*. Minneapolis, MN, University of Minnesota Press, 2005.

Connor, Steven, "An Air that Kills: A Familiar History of Poison Gas" (2003) http://www.bbk.ac.uk/english/skc/gas/

——. "Exhaust" lecture University of St. Andrews, June 16th 2006. http://www.bbk.ac.uk/english/skc/exhaust/

Corbin, Alain, *The Foul and the Fragrant: Odor and the French Social Imagination*. Translated by Miriam Kochab. Cambridge, MA, Harvard, 1986.

Crasset, Matali, *The Pigeon Loft*. Paris, Pyramyd, 2004.

Deleuze, Gilles and Guattari, Felix, *Kafka: Toward a Minor Literature*, Minneapolis, University of Minnesota Press, 1986.

——. *A Thousand Plateaus: Capitalism and Schizophrenia*. translated by Brian Massumi. Minneapolis, University of Minnesota Press, 1988.

Dessauce, Marc, *The Inflatable Moment: Pneumatics and Protest in '68*. New York: Princeton Architectural Press 1999.

Douglas, Mary, *Purity and Danger: An Analysis of Concepts of Pollution and Taboo*. London, Routledge, 2002.

Dumont, Gabrielle Pierre-Martin, *Les Ruines des Paestum*, 1764.

Elliott, C. D. *Technics and Architecture: The Development of Materials and Systems for Buildings*. Cambridge, Ma., The MIT Press, 1992.

Evans, Clinton, *The War on Weeds in the Prairie West: An Environmental History*. Calgary, Alberta Canada: University of Calgary Press, 2002.

Evans, Robin. *Translations from Drawing to Building and Other Essays*. London, Architectural Association Publishers, 1997.

Fathy, Hassan, *Architecture for the Poor: Am Experiment in Rural Egypt*. Chicago, University of Chicago Press, 1973.

Forty, Adrian, *Words and Buildings: A Vocabulary of Modern Architecture*. London, Thames and Hudson, 2000.

——. "Primitive, the word and concept" in Odgers, J., Samuel, F., Sharr, A. (ed.) *Primitive: Original Matters in Architecture*. London, Routledge (2006): 3-14.

Furlaud, Alice, "Homing in on Pigeon Towers" *New York Times*, April 1, 1990.

Gandy, Matthew, "The Paris sewers and the rationalization of urban space" *Transactions of the Institute of British Geographers* 24:1(1999): 23-44

Gessert, George. "Bastard Flowers" *Leonardo*, Vol. 29:4 (1996): 291-298.

Giedion, Siegfried. *Mechanization Takes Command.* Oxford, Oxford University Press, 1948.

Gilman, Sander and Zhou Xun eds., *Smoke: A Global History of Smoking*. London, Reaktion Books, 2004.

Gissen, David, ed. *Big and Green: Toward Sustainable Architecture in the 21st Century*. New York, Princeton Architectural Press, 2003.

——. "Exhaust and Territorialisation at the Washington Bridge Apartments, 1963-1973." The Journal of Architecture, Vol. 12:4 (2007): 449-461.

Goldhagen, Sarah Williams. "Freedom's Domiciles: Three Projects by Alison and Peter Smithson" in Sarah Williams Goldhagen and Réjean Legault, *Anxious Modernisms: Experimentation in Postwar Architectural Culture.* Cambridge, Ma, MIT Press, 2000.

Haraway, Donna. *Simians, Cyborgs and Women: The Reinvention of Nature*, New York and London, Routledge, 1991.

Harris, John, and Snodin, Michael, eds. *Sir William Chambers: Architect to George III*. New Haven, Yale University Press, 1996.

Harrison, Charles, Paul Wood, and Jason Gaiger. *Art in Theory, 1815-1900: An Anthology of Changing Ideas.* London: Blackwell, 1998.

Heaney, Seamus. *Bog Poems*, 1975.

Hersey, George. *The Lost Meaning of Classical Architecture.* Cambridge. Mit Press, 1988.

Hough, Michael, *Cities as Natural Processes: the Basis of Sustainability*. London, Routledge, 2005.

Hugo, Victor. *Les Misérables*, London, Penguin, 1982.

Inaba, Jeffrey. "Big Leaks," *Volume 5*, Archis, 2005

Isozaki, Arata. *Japan Architect #247*, Oct/Nov: 20-21, 1977.

Jencks, Charles. "Peter Eisenman: An Architectural Design Interview" in *Deconstruction, Architectural Design* (1988): 48-61.

Kessler, Marni Reva. *Sheer Presence: The Veil in Manet's Paris*, Minneapolis, University of Minnesota Press, 2006.

Kipnis, Jeffrey. Moonmark, Assemblage, 16 (1991): 6-13.

Knight, Richard Payne. *The Landscape*, 1794.

Koolhaas, Rem "'Life in the Metropolis' or 'The Culture of Congestion'" (1977) in Hays, K. Michael. *Architecture Theory Since 1968*. Cambridge, MA, MIT Press (1994): 320-331.

Kwinter, Sanford. "La Cittá Nuova: Modernity and Continuity" (1986) in Harry Francis Mallgrave. *Architectural Theory: An Anthology from 1871 to 2005.* London, Blackwell (2008): 474-475.

Laugier, Marc-Antoine. *An Essay on Architecture.* Trans. Wolfgang and Anni Herrmann. Los Angeles: Hennessey and Ingalls, 1750/1977.

Le Corbusier. *Towards a New Architecture*. New York, Dover, 1986.

——. *The City of Tomorrow and its Planning*. New York, Dover, 1987.

——. *Precisions: On the Present State of Architecture and City Planning*, Cambridge, Mit Press, 1991.

Le Roy, Julien David. *The Ruins of the Most Beautiful Monuments of Greece*, Los Angeles, Getty Publications, 1770/2004.

Levine, Neil, "The Significance of Facts: Mies' Collages Up Close and Personal." *Assemblage*, 37 (1998): 70-101

Lynn, Greg. "Body Matters," *Journal of Philosophy and the Visual Arts.* special issue on "The Body", edited by Andrew Benjamin. (1997): 61-69

Mallgrave, Hary Francis. *Architectural Theory, An Anthology from Vitruvius to 1870*. London, Blackwell, 2007.

——. *Modern Architectural Theory: A Historical Survey*, 1673-1968, Cambridge, Cambridge University Press, 2005

——. *Architectural Theory: An Anthology from 1871 to 2005.* London, Blackwell, 2008.

McCloy, Shelby T. "Flood Relief and Control in Eighteenth-Century France" *Journal of Modern History*, Vol. 13:1 (1941): 1-18.

Mertins, Detlef, *The Presence of Mies*. New York, Princeton Architectural Press, 1994.

Merwood, Joanna. "The Mechanization of Cladding: The Reliance Building and the Narratives of Modern Architecture", *Grey Room*, 4 (2001): 52-69.

Middleton, Robin and Watkin, David. *Neoclassical and 19th Century Architecture, Vol I & II*. New York, Rizzoli, 1980.

Miller, Naomi. *Heavenly Caves: Reflections on the Garden Grotto*. New York: George Braziller, 1982.

Mumford, Lewis. The Brown Decades: A Study of the Arts in America. New York, Dover, 1955.

Muschamp, Herbert. "Thank you for not smoking" *New York Times*, May 14, 1995.

Nead, Lynda. *Victorian Babylon: People, Images and Streets in Nineteenth-Century London*. New Haven, Yale University Press, 2000.

NL Architects, *Remix of Reality*, Seoul, DD series, 2005.

Nox (1999) http://www.archilab.org/public/1999/artistes/noxa01en.htm

O'Connor, Erin. *Raw Material: Producing Pathology in Victorian Culture*. Durham, Duke University Press, 2000.

Otero-Pailos, Jorge. Correspondence with the author, July, 2008.

Picon, Antoine. *French Architects and Engineers in the Age of Enlightenment.* Cambridge, Cambridge University Press, 1991.

——. "Nineteenth-Century Urban Cartography and the Scientific Ideal: The Case of Paris" *Osiris*, Vol. 18 (2003): 135-149.

——. Lecture at Anxious Landscape Symposium, University College London, 2005.

Porter, Dale. *Thames Embankment. Environment, Technology and Society in Victorian London, The.* Akron, OH: University of Akron Press, 1998.

Pugin, A. W. N. Contrasts: Or, A Parallel between the Noble Edifices of the Middle Ages and Corresponding Buildings of the Present Day. London, 1836.

Ramiréz, Juan Antonio, *The Beehive Metaphor: From Gaudí to Le Corbusier*. London, Reaktion, 1998.

Ranier, Roland. *Livable Environments*, Zurich, Verlag fur Architektur Artemis Zurich, 1972.

Read, Justin. "Alternative Functions: Oscar Niemeyer and the Poetics of Modernity." *Modernism/modernity* Vol. 12:2 (2005): 253-272.

Riani, Paolo. *Twentieth Century Masters: Kenzo Tange*. New York, Hamlyn, 1969.

Robbins, David. *The Independent Group: Postwar Britain and the Aesthetics of Plenty*. Cambridge, MA, The Mit Press, 1990.

Roth, Michael with Lyons, C and Merewhether, C. eds. *Irresistible Decay: Ruins Reclaimed*. Los Angeles, Getty Publications, 1996.

Ruskin, John. *The Stones of Venice, Second Volume.* London, Smith, Elder and Co, 1873.

——. *The Complete Works of John Ruskin, Volume I.* New York, Chicago: National Library Association, 1905.

——. *The Ethics of the Dust.* London, Merrill and Baker, 2003/1877.

Ryan, Raymund. *N3: The Irish Pavilion at the Seventh International Biennale, Venice*, Dublin, Ireland, Haus Publishing, 2001.

Rykwert, Joseph. *On Adam's House in Paradise*, Cambridge, Mit Press, 1997.

Schinkel, Karl Freidrich, David Bindman and Gottfried Riemann. *The English Journey: Journal of a Visit to France and Enland in 1826*. New Haven: Yale University Press, 1993.

Schnapp, Jeffrey. *Crowds.* Stanford, Stanford University Press, 2006.

Schumacher, Thomas L. *Surface and Symbol: Giuseppe Terragni and the Architecture of Italian Rationalism*, New York, Princeton Architectural Press, 1991

Scott, Felicity. *Architecture or Techno-Utopia: Politics after Modernism*. Cambridge, Mit Press, 2008.

Smith, Neil. *Uneven Development : Nature, Capital and the Production of Space*, London: Blackwell, 1991.

Smithson, Peter and Alison. *The Charged Void: Architecture, New York*, The Monacelli Press, 2001.

Steedman, Carolyn. *Dust: The Archive and Cultural History*. New Rutherford, Rutgers University Press, 2002.

Stern, D., Mellins, T and Fishman, R. *New York 1960: Architecture and Urbanism Between the Second World War and the Bicentennial*, New York: Monacelli Press, 1995.

Stoppani, Teresa. "Dust revolutions. Dust, *informe*, architecture (notes for a reading of Dust in Bataille)." *Journal of Architecture*, Vol. 12:4 (2007): 437-447.

Sullivan, Louis. *Kindergarten Chats and Other Writings.* Courthope Press, 2007.

Tait, A. A. "Reading the Ruins: Robert Adam and Piranesi in Rome," *Architectural History*, Vol 27 (1980): 524-533.

Tournikiotis, Panayotis. *The Historiography of Modern Architecture*, Cambridge, MA, MIT Press, 1999.

Vidler, Anthony. *The Architectural Uncanny: Essays in the Modern Unhomely*. Cambridge, MIT Press, 1992.

Viollet-le-Duc, Eugène Emmanuel, *Annals of a Fortress: Twenty-Two Centuries of Siege Warfare*. New York, Dover, 2007.

Vitruvius, Marcus Pollio. *The Ten Books of Architecture*. Trans. Morris H. Morgan. New York: Dover Publications, Inc., 1960.

Weingarden, Lauren S. "Naturalized Nationalism: A Ruskinian Discourse on the Search for an American Style of Architecture" *Winterthur Portfolio*, Vol. 24:1 (1989): 43-68.

Wendelken, Cherie. "Putting Metabolism Back in Place: The Making of a Radically Decontextualized Architecture in Japan." In Sarah Williams Goldhagen and Réjean Legault, *Anxious Modernisms: Experimentation in Postwar Architectural Culture*. Cambridge, MA, MIT Press, 2000.

Woodbridge, S. H. *Upwards versus Downwards Ventilation*. London, Robert Boyle and son, 1900.

Wolf, Peter. *The Evolving City: Urban Design Proposals by Ulrich Franzen and Paul Rudolph*. New York: The American Federation of the Arts, 1974.

Wright, Frank Lloyd. *An Autobiography*. New York, Duell, Sloan and Pearce, 1943.

INDEX

Page numbers in italics indicate images

IMAGE CREDITS

All images of contemporary architecture (1978-present) are provided courtesy of the architects unless otherwise noted.

Fig. 1.1: Johns Hopkins University, Fowler Collection
Fig. 1.2: Bibliotheque Nationale de France
Fig. 1.3: Getty Images
Fig. 1.4: Fondation Le Corbusier/Artists Rights Society/ADAGP, Paris
Fig. 2.1: Johns Hopkins University, Fowler Collection
Fig. 2.2: Corbis
Fig. 2.3: Bildarchiv Preussischer Kulturbesitz/ Art Resource
Fig. 2.4: Johns Hopkins University, Fowler Collection
Fig. 2.5: Getty Images
Fig. 2.6: The Art Institute of Chicago, Chicago, Ryerson and Burnham Archives, Historic Architecture and Landscape Image Collection
Fig. 2.7: Getty Images
Fig. 2.8: Bill Maris/ESTO
Fig. 2.9 (background photograph): Andrew Kudless
Fig. 3.2: Giraudon Collection/ Art Resource, NY
Fig. 3.3: Getty Images
Fig. 3.4: Berkeley Art Museum, University of California, Berkeley, courtesy Chip Lord
Fig. 3.7: 2x4, New York
Fig. 4.1: Fondation Le Corbusier/Artists Rights Society/ADAGP, Paris
Fig. 4.2: New York Public Library
Fig. 4.3: The Museum of Modern Art
Figs. 4.4 + 4.5: Getty Images
Fig. 5.2: Bibliotheque Nationale de France
Figs. 5.3 + 5.4: Getty Images
Figs. 5.7 + 5.8: Cameraphoto Arte/La Biennale di Venezia
Fig. 6.1: Johns Hopkins University, Fowler Collection
Fig. 6.3: Bibliotheque Nationale de France
Fig. 6.4: The Library of Congress, Washington, DC
Fig. 6.5: Corbis
Fig. 6.6: The MIT Press, Cambridge, MA
Fig. 6.7: Royal Institute of British Architects, London
Figs. 6.8 - 6.10: Daria Scagliola
Fig. 6.18: Theodor Rzad/DigiTEd
Fig. 7.1: Johns Hopkins University, Fowler Collection
Fig. 7.2: Corbis
Figs. 7.3 + 7.4: Getty Images
Fig. 7.5: Corbis
Fig. 7.6: Royal Institute of British Architects, London
Figs. 7.7 - 7.10: Dennis Mortell
Fig. 8.1: Johns Hopkins University, Fowler Collection
Fig. 8.3: Corbis
Fig. 8.4: Tate Gallery, London/ Arts Resource, NY
Fig. 8.5: Royal Institute of British Architects, London
Fig. 8.6: Getty Images
Fig. 8.7: The Museum of Modern Art
Fig. 9.1: The Art Institute of Chicago, Ryerson and Burnham Archives, Chicago Architectural Sketch Club Collection
Fig. 9.4: The Art Institute of Chicago, Chicago, Ryerson and Burnham Archives, Chicago Architectural Sketch Club Collection
Fig. 10.1: New York Public Library
Fig. 10.2: Getty Images
Fig. 10.3: Frank Lloyd Wright Foundation/ Artists Rights Society (ARS), NY
Fig. 10.4: New York Public Library
Fig. 11.1: New York Public Library
Fig. 11.4: Fondation Le Corbusier/Artists Rights Society (ARS)/ADAGP, Paris
Fig. 11.5: Kristina Seier
Figs. 11.12 + 11.3: Elizabeth Felicella
Fig. 12.1: Institute of Civil Engineers, London
Fig. 12.3: The Museum of Modern Art, Artists Rights Society (ARS), New York/VG Bild-Kunst, Bonn
Fig. 12.4: Estate of R. Buckminster Fuller